Energy Planning and Policy

Energy Planning and Policy

The Political Economy of Project Independence

Thomas H. Tietenberg
Williams College

With the assistance of
Pierre Toureille

Lexington Books
D.C. Heath and Company
Lexington, Massachusetts
Toronto   London

Library of Congress Cataloging in Publication Data

Tietenberg, T.H.
  Energy planning and policy.

  Includes index.
  1. Energy policy—United States. I. Toureille, Pierre, joint author.
II. Title.
HD9502.U52T53   333.7   75-13309
ISBN 0-669-00048-5

Copyright © 1976 by D.C. Heath and Company.

Published simultaneously in Canada.

Printed in the United States of America.

International Standard Book Number: 0-669-00048-5

Library of Congress Catalog Card Number: 75-13309

**To Gretchen and our children,
Heidi and Eric**

Contents

List of Tables

Preface

An increasing number of widely read works by respected scholars in a number of disciplines are putting forth and defending the thesis that the human prospect is not particularly bright. The source of concern can frequently, if not always, be traced to a conviction that the political and economic systems, which have been erected to govern human relationships, are inadequate to meet the challenges that will confront them in the coming decades, much less in the coming centuries.

Project Independence, President Nixon's highly publicized attempt to erect a comprehensive and coherent energy policy, affords a unique opportunity to examine this general thesis within the context of current domestic history. Since the issues involved were highly visible and there were no easy ways out, it represents an interesting test case for assessing the responsiveness of political and economic institutions.

Recognizing the importance and complexity of the issues, the government established and funded an analytical effort of unprecedented magnitude as an integral part of Project Independence. Conceived as an analytical fusion of the natural and social sciences, this effort involved some 2,000 persons either part time or full time in providing inputs for and in analyzing outputs from a single integrated modeling system. It was, in short, an impressive attempt to deal with the energy problem by marshalling the available evidence, evaluating it, and translating the results into a comprehensive and coherent policy framework.

This book documents what went on within the government during the first two years of Project Independence and extracts from this discussion the forces that affected the responsiveness of the policy process. It characterizes the historical precedent, describes the response taken to quantify the consequences of alternative actions, discusses the integration of these results with political considerations to form the energy packages that emerged, and explains the reasons for the metamorphosis of each of these packages as they proceeded through the policy process. The final section provides an evaluation of the substance and process of energy policy. In Chapter 9 a decision theory model, which characterizes possible future states of the world and quantifies the relationships among current policies and future outcomes, is used to describe the desirability of a range of policy packages. The final chapter suggests what can be learned from this case study about the responsiveness of the policy process.

Since the policy process never really ends, but continually refines and updates the policy framework, a book on the subject can never be completed unless the analysis is restricted to a particular historical period. In this book the analysis is restricted to the period prior to October 31, 1975. Particularly intensive coverage is given to the two-year period following the 1973 Arab oil embargo. This is a historically significant end point both because it represents the end of a

xiii

full two-year search for a comprehensive energy policy and because it was the end of the Arab agreement to freeze oil prices. In terms of negotiation it would have been helpful to have the rudiments of a United States policy firmly in place by that date. In fact they were not, but that is getting ahead of the story.

As a Brookings Economic Policy Fellow I was afforded a somewhat unique vantage point for writing this book. By the time I arrived at the Federal Energy Administration (late June 1974) the modeling process was well under way and the major decisions shaping the basic structure of the modeling system had already been made. As a result I have no stake in those decisions and can view them in retrospect from my teaching position with a certain detachment. On the other hand, as a division director I was immediately drawn heavily into the Project Independence modeling effort and was keenly aware of its uses and abuses during the policy formulation process.

In writing this book I have been aided by many persons. I am particularly indebted to Pierre Toureille, my research assistant. In addition to insuring that the more tedious but important details of publishing a book were accomplished, he made significant substantive contributions to Chapters 7 and 8. In addition I owe a debt of gratitude to my many colleagues at the Federal Energy Administration, the Brookings Institution, and Williams College, who shared their time and insights with me. I wish particularly to acknowledge the valuable comments made by Mike Brazzel, Hill Huntington, Bill Hogan, Bart Holaday, Bruce Pasternack, Mark Rodekohr, Al Cook, and Steve Chapel at the Federal Energy Administration (FEA); James Sundquist, Charles Schultze, George Perry, and Nina Cornell at the Brookings Institution; and MacAlister Brown and Tom Jorling at Williams College. In addition, I wish to thank members of the support staff at the three institutions. Ozzie Simms of the FEA Library and both Edith Kirby and Laura Walker of the Brookings Library were especially helpful in tracking down illusive documents. The typing of various drafts was expertly accomplished by Janet Murrill of FEA, Valerie Harris of the Brookings Institution, and Constance Ellis at Williams College. Marcia Appel of the Brookings Institution was extremely helpful in providing me with a productive research environment during my residence at the Brookings Institution. The strengths of the book reflect these helpful contributions but the weaknesses are my own. I am grateful to all.

Energy Planning and Policy

1 Introduction and Overview

Objectives of the Book

In an important sense the landscape of human events is moving more rapidly now than before. Exponential growth implies that each additional increment to GNP, to population, to knowledge, etc., is larger than the last. Several scholars,[1] coming from widely different intellectual traditions, have recently concluded that the ability of persons and institutions to cope with and adapt to this accelerating pace of activity is likely to be quite limited.

The costs of nonadaptation can be very large. In its highly controversial study, *The Limits to Growth*,[2] the Club of Rome portrays a future in which the world exhausts its supplies of vital resources and the result is a rather precipitous collapse of society as we now know it. One of the main themes of the report is that an inability to adapt to conditions of increasing scarcity would cause a sudden and abrupt halt to economic activity.

Critics of the report[3] quarrel less with the conclusions concerning the costs of nonadaptability than with the issue of whether man and his institutions can adapt. The Club of Rome study, these critics argue, fails to recognize the existence and importance of an adaptation mechanism that would automatically restructure incentives and thereby modify behavior in ways to prevent the collapse.[a] This mechanism, the price system, would stimulate the discovery of new resources, would lead to the development of processes that could substitute more abundant resources for the scarce ones, would cause consumers to conserve, would make recycling profitable, and on and on. History is full of examples of how this mechanism has served this function and served it well.[4] By implication, it should be expected to continue to fill this role in the future.

As appealing and comforting as this argument is, it is somewhat incomplete, for it neglects the significant fact that the price system operates within a rule of law. The legal environment, within which the price system works, places important constraints on the market. It removes much of the automatic character of the price system. When the government is an active participant in the market process, both as a buyer and a seller of resources and as a regulator of other participants, the responsiveness of the price system to conditions of increasing scarcity depends on the responsiveness of the policy-making political

---

[a]Not all economic models lead to this conclusion; see Michael Common and Daniel Pearce, "Adaptive Mechanisms, Growth, and the Environment: The Case of Natural Resources," *The Canadian Journal of Economics* VI (August 1973): 290-300.

1

institutions. Their adaptability and responsiveness to conditions of increasing scarcity are much less clear.

There are two significant characteristics of energy problems that cause concern about the responsiveness of our political institutions: First is the complexity of the factual base on which the issues rest. How injurious to health is a particular air pollutant? How will demand react to a tax? How important will foreign suppliers be in 1985? How much substitutability exists in the use of the scarce resource? At what prices does this substitution become economically viable? How do various policy options affect different categories of voters? Informed choice requires gathering, structuring, interpreting, and evaluating massive amounts of information, as well as developing a means for dealing with issues for which information is unavailable. For the available information to be brought to bear on the problem, effective information channels are required and politicians must be willing to engage in long-range planning, even when current costs are more visible to the electorate than the future benefits to be derived. The ability of elected officials to engage in long-range planning, when their tenure in office may be limited to from two to six years, depending on the office held, is not obvious.

The second major cause for concern is that energy policy choices involve direct conflicts among important national goals. Conservation can cost jobs. Exploiting new resources can lead to conflict with environmental goals. Allowing prices to rise, the modus operandi of the price system, causes concern over inflation and the equitability of these price increases on the disadvantaged members of society. Higher prices are visable to voters, as are the generally higher profits received by owners of the depletable resource. The path of least resistance in confronting these volatile political issues may be inaction or action that specifically retards the responsiveness of the economic system.

Project Independence, President Nixon's highly publicized program to achieve energy self-sufficiency by the 1980s, represents the most recent and, in many ways, the most important attempt by the federal government to grapple with these problems. It contains all the dimensions that challenge the responsiveness of democratic institutions. At stake was the use of two valuable depletable resources—oil and natural gas. The choices involved direct conflicts between societal goals. The energy system was, and is, exceedingly complex, involving a number of products, many participants, and an important geographic dimension. The vested interests in particular choices were pronounced.

Because these were highly visible and important issues, a serious attempt to provide a policy framework was undertaken by the policy-making institutions. The Project Independence analysis was one of the largest analytic efforts ever undertaken in government. It combined the talents of literally hundreds of natural and social scientists, who worked together to build a single, integrated modeling system. A mass of information was collected, synthesized, and analyzed for the policy makers. Alternatives were weighed and decisions reached. This book represents an attempt to record, to characterize, and to evaluate this response.

The book examines both the substance and process of energy policy formation. Several questions guide the inquiry: How responsive was the policy-making institutional structure in the executive and legislative branches to the need for a comprehensive policy framework? What comprehensive planning models were developed to facilitate policy making and how were they integrated into the process?[b] What factors shaped the substance of the policy as it proceeded through the various stages of the process? How well did the substance of the policy, after having been shaped by these forces, fulfill the initial objectives? What possible generalizations or hypotheses about the responsiveness of governmental policy-making institutions can be gleaned from this important milestone in American history?

The effectiveness of the flow of information from the analyst to the policy maker is limited (or enhanced) by several factors, including the incentive structure that both participants face. Since a major portion of this book is concerned with the role of comprehensive planning models in the policy process, the next few parts explore the nature of the analyst-decision-maker relationship and its implications for the policy process.

The Role of Analysis in the Policy Process

A case study, such as this one, should be conducted within the context of the existing literature. The literature provides a point of departure and a sense of perspective to the case study. The case study provides an opportunity to subject theory to practice, and, more importantly, to develop new hypotheses. An exhaustive survey of the literature, however, is neither necessary nor desirable for our purposes because the range of possibilities is adequately suggested by reference to a couple major positions.[c] Our interest, of course, is in deriving the role of analyses for each model of the policy process.

---

[b]The role of planning models in formulating policy is a timely concern, quite apart from the importance of the subject matter these models address. There is an increasing interest in the incorporation of economic planning into the policy process by some politicians. Worth noting, for example, is S-1795, "The Balanced Growth and Economic Planning Act," initiated by Senators Humphrey and Javitts, which would set up an economic planning function to guide the policy process toward overall goals. The broad outlines of this bill and the rationale behind it are described in Myron E. Sharpe, "The Planning Bill," *Challenge* XVIII (May/June 1975): 3-8. The other attempts by Congress to integrate comprehensive planning into the policy process are discussed in Constance Holden, "Futurism: Gaining a Toehold on Public Policy," *Science* 11 July 1975, pp. 120-22, 124.

[c]Two standard references are Charles E. Lindblom, *The Policy-Making Process* (Englewood Cliffs, New Jersey: Prentice-Hall, 1968), and Charles L. Schultze, *The Politics and Economics of Public Spending* (Washington: Brookings Institution, 1968). A somewhat less standard work, which deals with the potential for the expert's values to dominate the process, is Guy Benveniste, *The Politics of Expertise* (San Francisco: Boyd and Fraser for the Glendessary Press, 1972).

The objective here is to draw upon the existing literature to provide a simple, positive[d] model of the decision-making process, which will be used as a framework for interpreting the process as it actually occurred. The development of such a perspective is important because it permits the integration of chronologically and institutionally separated events into general themes. The cost of not imposing this kind of framework is, frequently, an inability to assess the meaning of a litany of events in the policy process, events which are chronologically ordered, but otherwise apparently unrelated.

The mode of presentation that follows illustrates the spectrum of decision-maker-analyst roles, which is theoretically possible by reference to a couple of simple, rather extreme, models. The middle ground of this spectrum is then explored with the assistance of a succinct verbal model of the policy process. This model postulates certain objectives for decision makers and analysts and derives the nature of their relationship from these objectives. Since the purpose of this model is to provide a perspective for thinking about energy policy, no claim is made for its universality or completeness.[e]

At one end of the spectrum defining decision-maker-analyst roles is the *analyst-utopian model.* In this model a central authority, which is fully cognizant of the needs and desires of its citizenry, chooses those policy instruments that maximize social welfare. The analyst, in this model, is indispensable, since it is he who must derive the social optimum solution. In this case, the decisions are made de facto by the analyst, and the decision-maker is little more than a figurehead. It is clear why this model carries a lot of appeal in the analytical disciplines; it makes the analyst the dominant figure.[f]

At the other end of the extreme is the *politician-utopian model.* In this model information is not available, uncertainty is rampant, and the objectives for the policy are not amenable to quantification. The role of the analyst in this kind of environment is minimal or nonexistent. The absence of key relevant information makes scientific analysis impossible. The presence of uncertainty may require subjective judgments about the relative likelihood of alternative events. When the objectives are inherently nonquantifiable, the quantification of the relationship between ends and means is impossible. The decision makers, in this model, are relatively unconstrained in their choices by analysis.

Somewhere in between these two caricatures lies a reasonably accurate representation of the executive branch decision-making process and the role of

---

[d]The word positive here means descriptive. Positive models can be contrasted with normative models, which describe how the policy process should work.

[e]For example, the model has limited applicability to congressional decision making although it yields some important insights on aspects shared with the executive branch. Models of legislative decision making in general lag behind executive models. See John A. Ferejohn and Morris P. Fiorina, "Purposive Models of Legislative Behavior," *The American Economic Review* LXV (May 1975): 407-14.

[f]The importance of this view in economics is a case in point. The implications of this are further explored in James M. Buchanan, "A Contractural Paradigm for Applying Economic Theory," *The American Economic Review* LXV (May 1975): 226.

the analyst. In attempting to model this middle ground we assume for simplicity, the existence of a single decision maker and a single analyst. The decision maker · is assumed to be interested in being reelected. His decisions are undertaken to maximize the number of votes he will receive in the next election. The basic problem for the decision maker is that he has to make decisions on a wide variety of subjects and for a good number of these he generally has no particular expertise; it is therefore unclear what the consequences of a particular decision are.

This provides a need for analysts of all kinds—political analysts, economic analysts, natural scientists, etc.—to collect the relevant, available information and systematize it, so that it provides the quantity and quality of information needed to reach a decision within the constraints imposed on the analysis.

The second participant in this process, the analyst (or, more generally, the manager of the analytical function) is assumed to be guided by his desire to maximize his budget allocation. The more resources he has at his disposal the more information he can provide. To the extent that information conveys power, then this assumed objective is consistent with a possible alternative objective, the desire to accumulate personal power.

These two behavioral assumptions lead to a variety of hypotheses about the politician-analyst[g] relationship. Analysis requires a condensation of information. This requires value judgments by the analysts as to what is and what is not important. Each such judgment reduces the prerogatives of the decision maker. The relationship between analyst and decision maker, therefore, is potentially one of conflict. Over time this conflict is resolved through positive and negative reinforcement. When the analyst makes appropriate choices, his analysis is used and his budget requests are reviewed favorably by the politician. When the analyst makes inappropriate choices, his analysis is ignored; he has no impact. Since the analyst wants to protect his jurisdiction and maximize his budget, he learns to be useful.

Each decision maker approaches his function with a particular perspective or ideology.[h] This ideology consists of a set of principles about the world and the way it operates. These principles are accepted as valid by the decision maker and tend to establish a framework for looking at policy decisions. For example, in the case of the last two Republican administrations this ideology would include a faith in the market as an allocation mechanism, a belief in minimizing the role of government and a conviction that the way to improve the economic well-being of the voters is through promoting business prosperity.[i]

---

[g]Politician and decision maker are used interchangeably in what follows because the decision maker has been assumed to be an elected official who is motivated by his desire to be reelected.

[h]Ideology, of course, is not restricted to the decision maker. For a discussion of the ideology of economists and its root causes, see Duncan Foley, "Problems as Conflicts: Economic Theory and Ideology," *The American Economic Review* LXV (May 1975): 234.

[i]An articulation of this view can be found in William Simon, "Getting Government Out of the Marketplace," *Saturday Review*, 12 July 1975, pp. 10-13, 16, and 20.

Ideology affects the relationship between analysis and decision making by constraining the range of alternatives considered.[j] Alternatives clearly contrary to the decision maker's ideology have a lower probability of being accepted than alternatives that are compatible with this ideology. Therefore, to maximize impact for a given amount of effort, the analyst usually stays within the ideological framework that provides the lens through which the decision maker sees the world.[k]

The analyst can provide services to the decision maker at three points in the decision-making process: He can suggest policy alternatives. He can provide valuable assistance in policy selection by clarifying the relationships among policy choices and objectives. His analysis can also provide a valuable role in helping the politician justify the policy, once implemented. These roles are not independent. One of the criteria that would likely be used by the politician in selecting a policy is how well it can be justified and explained to the voters. The President, for example, as one of his devices to prod a reluctant Congress into action, frequently relies on the technique of selling his program to the voters through various, well-publicized speeches. The program, if it is to be successfully sold, must have an appealing logic behind it and must be able to withstand attacks from the opposition party.

Having spelled out the potential role that can be played by analysis it is now possible to discuss two principles that govern how much of this potential will be realized in any particular politician-analyst relationship. First, in order to be used, the analysis must be timely. The political landscape changes rapidly; waiting until all the facts are in is frequently not a politically viable alternative. If the analysis is complete when the time for decision has arrived, it can be used. If not, the decision will be made without it.

The second principle is that the analysis has to be controllable. This does not necessarily mean that the outcomes have to be preordained, although, in some cases, it means exactly that. It does mean that the decision maker has to be able to define the ground rules under which the study will be conducted. This includes specifying the objectives, agreeing with the assumptions, and participating in the specification of the range of alternatives considered.

The principle of timeliness can severely limit the role of analysis because of a time compression phenomenon that occurs in bureaucracies. There are usually several levels of bureaucratic hierarchy between the analyst and the politician, particularly when the decisions involve long-range planning. Consider, for

---

[j]Ideology is, of course, not the only factor limiting the range of alternatives considered. As Lindbloom has persuasively argued, previous policy is rarely reexamined and, therefore, only alternatives in addition to this historically defined package may be actively considered. See Charles E. Lindbloom, *The Policy-Making Process* (Englewood Cliffs, New Jersey: Prentice Hall, 1968).

[k]A similar role for ideology in decision making and an investigation of its implications for foreign policy can be found in Morton H. Halperin, *Bureaucratic Politics and Foreign Policy* (Washington: Brookings Institution, 1974).

example, the following hypothetical sequence of events: The President identifies an emerging problem. He then requests information from a department head. The department head gives it to his executive communications expert who gives it a control number, decides to whom the task should be assigned, and puts it in the intraagency mail. To insure that the department head has a chance to see, comprehend, and pass judgment on the information, an intradepartment deadline is set at least two or three days prior to the White House deadline. Similar deadlines are set by each lower level of the hierarchy. By the time the analyst receives the question a good deal of time has already elapsed and, as a result, he has only a short time to prepare a reply. The net result of this time compression phenomenon may be that the President may receive his answer in a month but the analyst was given only two days to prepare the answer. The readers of this book who have performed in an analytical role in Washington will recognize that this is not an uncommon phenomenon.

In general, then, this simple model suggests several aspects of the role of analysis in policy that should be kept in mind when reading this book: At what level in the hierarchy was the analysis accomplished? How did this affect controllability and timeliness? Did the analysis successfully satisfy the potentially serious conflict over the preemption of the prerogatives of the decision maker? If so, how? What were the constraints on the analysis? What was the relationship between the decision makers and the analytical structure in dealing with uncertainty? Was the analysis used in the policy selection process or merely to justify preordained decisions?

Comprehensive Planning Models as a Special Case

The foregoing model describes the relationship between a politician and an analyst in very abstract terms. Of course, the term *analyst* covers a lot of territory, ranging from someone performing back-of-the-envelope calculations to a person designing and using complex simulation models. Since our focus in this book is on the role of comprehensive planning models, it is necessary to apply the principles of the politician-analyst relationship to a situation in which the use of such models is at stake.

Why are comprehensive policy impact models needed? Comprehensive planning models are needed to relate present policy alternatives to future outcomes in a way that accounts for the positive and negative feedbacks which occur among different policies and among different sectors of the economy. History is full of examples of sequential decision making in which each decision appears perfectly rational, when isolated in time or space, but, when the full set of policies is examined over the full period of history, glaring deficiencies appear. Indeed, energy policy provides a classic example, as shown in Chapter 2.

This argument is not new. It has arisen, for example, in the debate about federal expenditure and taxation policies and the goal of stabilizing the economy. The federal budget is an aggregate of individual revenues and expenditures. In a given budget each line item may be perfectly justified in terms of the individual goals of that program, but the total budget can still be inconsistent with the aggregate goal of economic stabilization. Currently there are several large-scale macroeconometric models in use that try to perform this comprehensive planning function by relating the package of individual policies to the goal of economic stabilization. They allow the budget forecaster to study the impacts of certain tax and expenditure decisions on labor markets, inflation, gross national product, the housing market, etc. These impacts are internally consistent and consistent with history, since the equation systems on which they are based were estimated from historical data.

The analogy is perfectly applicable to energy policy. Comprehensive planning models are needed to assess alternative energy policy packages because the goals are primarily aggregate in nature and each policy package is likely to be accompanied by significant indirect effects, which will tend to enhance or diminish the achievement of these aggregate goals.

Unfortunately the situations that call for comprehensive planning models are precisely the situations in which the politician-analyst relationship is most strained. This can be illustrated by applying the principles of timeliness and controllability.

Timeliness causes a particularly acute strain on the relationship when comprehensive planning models are involved. These models, because they are designed to capture the simultaneous interaction of a large number of variables and the feedbacks on the system as a whole, are inherently complex. This complexity translates into long development periods and large setup costs. In addition, it usually leads to slow turnaround times between the time of problem definition and the time when the model solution is presented to the politician in terms he can understand and use. The politician, who normally needs rapid and well articulated input, will not, in general, be sympathetic to the many, very real and very important problems that crop up to make timeliness an illusive goal.

Controllability is also more of a problem for these modeling efforts than for simpler analytical functions and this tends to diminish their influence as well. The models are generally so complex as to be fully understood only by their developers and a relatively few, very experienced users. The number of inputs is huge and the impact of any particular input on the final solution is frequently unclear. Sensitivity analysis can be performed, but the accomplishment and comprehension of a complete sensitivity analysis is a practical impossibility because of the large number of variable combinations involved. As a result the politician can never hope to exercise complete control over all the assumptions and inputs; these must be left to the analyst.

Thus, whether a comprehensive modeling effort will have an impact on policy depends on the extent to which the intrinsic problems of timeliness and controllability can be solved. It also depends on such factors as the clarity of the objectives, the quality and quantity of available input information, and the intuitiveness of the results to the politician. As one recent case study on the role of a large-scale model on pollution control decision making for the Delaware River Basin found, the effective use of large-scale models in policy making is far from a foregone conclusion.[5]

An Overview of the Book

With the aid of this perspective we now turn to an examination of the substance and process of energy policy. Part I provides a historical context for the choices to be made. What was the historical precedent? How did we get where we are today? What forces shaped this historical evolution and what are their implications for future policy? What was the background of the Arab oil embargo of 1973? What was the significance of the embargo?

Part II describes the analytical response undertaken to assess the feasibility and desirability of alternative courses of action. It presents the conceptualization of the problem that guided the analysis and delineates the constraints that limited its usefulness. It describes the analytical system used and the results derived from it.

Part III discusses the legislative and executive policy processes as they grappled with the formulation of energy policy. It is a story richly endowed with conflict and compromise, bureaucratic politics and internal power struggles, and some institutional intransigence as well as some adaptability.

Part IV is devoted to an evaluation of the policy choices and the policy process. The information generated by the Project Independence modeling system is blended with some basic economic reasoning in a decision theory model to assess the costs of the various packages that have appeared during various stages of the policy process. The book concludes by generalizing from the observations made in the previous chapters to form some specific hypotheses about the complementarity and responsiveness of political and economic institutions to the need for managing the nation's depletable resources.

Notes

1. Alvin Toffler, *Future Shock* (New York: Random House, 1970), and Robert L. Heilbroner, *An Inquiry into the Human Prospect* (New York: W.W. Norton, 1974).

2. D.H. Meadows et al., *The Limits to Growth: A Report for the Club of Rome's Project on the Predicament of Mankind* (New York: Universe Books for Potomic Associates, 1972).

3. Carl Kaysen, "The Computer That Printed Out W*O*L*F," *Foreign Affairs* L (July 1972): 660-68; Scott Gordon, "Today's Apocalypses and Yesterday's," *The American Economic Review* LXIII (May 1973): 106-10; Robert M. Solow, "Is the End of the World at Hand?", *Challenge* XVI (March/April 1973): 39-50.

4. Glenn Hueckel, "A Historical Approach to Future Economic Growth," *Science* 14 March 1975, 925-31; Nathan Rosenberg, "Innovative Responses to Material Shortages," *The American Economic Review* LXIII (May 1973): 111-18.

5. Bruce A. Ackerman et al., *The Uncertain Search for Environmental Quality*, (New York: Free Press, 1973).

**Part I:
The Setting**

2 The Energy Situation in Historical Perspective

Although the high degree of visibility that the energy situation currently enjoys is a recent phenomenon, the forces that led to this situation were initiated some time ago. They represent a complex tangle of market forces, government policy, and international situations, which, when their full interactions had been felt, created a rather unfortunate energy situation for the United States. In a sense this situation represents a classic case of the failure of piecemeal, sequential policy, because, while each policy had a certain logic behind it in terms of the limited objective it attempted to satisfy, taken as a whole, the package was rather devastating.

The Increasing Demand for Energy

The logical place to begin this story is with a discussion of the historical demand for energy in the United States and the forces that led to this particular demand pattern. Although, in general, it is misleading to discuss energy as a homogeneous commodity, because different forms of energy are not perfectly substitutable for each other, such a discussion does have pedagogic merit in explaining aggregate trends in energy consumption over time. Therefore, we begin by concentrating on energy as a composite commodity, although individual fuels are discussed later in the chapter.

Total energy consumption can be thought of as being made up of two components: The first is the energy that is consumed directly by households, industry, and the commercial and transportation sectors. The second component is the energy that is lost in extracting, converting, transmitting, and consuming the various energy sources. The largest measurable source of these losses is the conversion of primary fuels into electric power and the transmission of this power from the point of production to the point of use. Table 2-1 presents some key growth rates that are helpful in understanding the recent history of these two components of energy growth. Since energy consumption grew faster than the population, it is not possible to explain the magnitude of this growth merely by attributing it to an increase in the number of consumers. Each consumer, on average, was consuming an increasing amount of energy each year. Two factors were instrumental in causing energy consumption per person to rise as rapidly as it did: rising incomes and falling relative prices of energy.

During the ten years from 1962 to 1972 the wholesale price index for all

Table 2-1

Annualized Growth Rates for BTU Energy Consumption in the United States by End Use Category and Selected Comparative Growth Rates 1950-72

	Annual Average Growth Rates	
Variable	1950-72	1962-72
Total domestic consumption of energy	3.5%	4.3%
Total electric utility conversion and transmission losses	5.6	7.4
Total end use consumption of energy	3.1	3.7
Transportation sector	3.3	4.3
Industrial sector	2.6	3.3
Household and commercial	3.7	3.8
Real GNP (1958 dollars)	3.7	4.1
Population	1.5	1.1

Source: The BTU growth rates were computed from tables H-1 through H-8 in Federal Energy Administration, *Project Independence Report* (Washington: U.S. Government Printing Office, 1974), appendix pages 9-16. The 1958 constant dollar GNP growth rates and the population growth rates were calculated from data in the U.S. Department of Commerce, *Business Statistics: The Biennial Supplement to the Survey of Current Business* (Washington: U.S. Government Printing Office, 1973), pp. 4 and 68 respectively.

Note: BTU refers to the British Thermal Unit, a convenient and widely used measure for comparing the energy content of different fuels. It represents the amount of heat required to raise the temperature of one pound of water from 62° F to 63° F.

commodities increased an average of 2.3 percent per year[1] while the wholesale price index for fuels and power rose only an average of 2.0 percent per year.[2] During this same period, energy prices also rose more slowly than the unit cost of labor.[3] Thus, while the absolute price of energy was rising over the period, the relative price was falling. The GNP growth rates combined with the population growth rates in Table 2-1 also indicate that this was a period of rising real per capita incomes.

These income and price effects were mutually reinforcing. Falling relative prices of energy increased energy consumption in two different ways. On the one hand it led to changes in the production process itself. Energy became an increasingly attractive substitute for other inputs, principally labor. The available econometric evidence suggests that the substitution of capital and energy for labor was quite price sensitive.[4] On the other hand this substitution prevented the costs of production from rising as much as they would have otherwise if energy had not been substituted for labor, causing a shift in relative output prices in favor of energy intensive products. As a result, in response to the lower relative prices of energy intensive commodities, consumers began to shift the allocation of their budgets away from labor-intensive commodities and toward energy-intensive ones. This, of course, merely intensified the rise in energy consumption initiated by the initial change in production techniques and the rise in income.

Rising incomes were yet another reason for rising energy consumption. A study conducted in 1973 on energy consumption patterns discovered that consumer expenditures on the direct consumption of energy (e.g., oil for space heating, gasoline for cars, electricity for appliances) were higher for high-income people than for lower income people, but lower income people spent a larger percentage of their budget on the direct consumption of energy.[5] This tells only part of the story, however, because a large amount of energy is also indirectly consumed. The production and transportation of nonenergy products takes energy so changes in the consumption of these goods affect the level of energy consumed. A study relying on 1963 data[6] concluded that when this indirect consumption of energy was taken into account, the total amount of energy consumed rose even more dramatically with income, since higher income groups consume substantially more indirect energy than the poor. The increase was still less than proportional, however, implying that the elasticity of energy consumption with respect to income is less than 1.0.

The influence of prices and incomes on energy consumption is of fundamental importance in assessing future energy policies. Since the energy-GNP ratio and the energy-population ratio have been influenced by prices, this is evidence of the existence of substitution possibilities between energy and other commodities. This implies that price-based conservation measures can reduce consumption without a proportional reduction in GNP. The fact that energy consumption rises somewhat more slowly than income implies that with a constant relative price of energy the rate of growth of real GNP should exceed the rate of growth of energy consumption.

As compelling as this argument is it has to be tempered by two caveats. The ability to reduce energy consumption without adversely affecting production is much greater over the long run than the short run. Capital stocks and productive processes are not instantaneously malleable; the changes usually come from replacing old energy intensive capital stocks with new, less energy-intensive ones. This replacement process tends to occur rather slowly. In the short run the opportunities to reduce energy consumption in the industrial sector, without reducing production, are rather more limited.

The second caveat concerns the reversibility of the substitution process. Even though lower relative prices led to a change in the way production and consumption activities were organized, this does not necessarily imply that price increases would cause as dramatic a change in the other direction. After the initial decisions are made, certain rigidities set in, which limit future flexibility. The importance of these two caveats is underscored by a more detailed look at the changes that took place in the transportation sector. Similar stories can be told about other sectors.

The large increase in energy consumption in the transportation sector was the result of increases in the number of vehicles being driven, increases in the number of miles driven, and increases in the amount of energy consumed per mile driven. These situations in turn resulted from an interlocking series of decisions, which were affected by relative prices and incomes, decisions concerning the mode of travel and residential choice.

A major theory of urban economics, which has been supported by empirical testing, suggests that consumers trade off transportation costs and land rents when choosing their place of residence.[a] One of the implications of this theory is that a reduction in transportation costs will make living further away from the workplace more attractive. When one couples this theory with the empirical observation that relative energy prices have been falling, it leads to the conclusion that energy prices probably contributed to the suburbanization of our cities during the last few decades.

This suburbanization trend, in turn, had important effects on the mode of travel chosen by the suburban population. As the people moved into the suburbs, new high-speed highways were built to provide better access to central city employment locations. These highways lowered travel costs for automobile travel relative to rail travel and caused some switching from trains to autos. In addition, suburbanization led to a reduction in the residential density. Since the economically efficient operation of mass transit systems (especially rail, but also, to a lesser extent, buses) requires high-density travel corridors, this also reduced the competitiveness of mass transit systems.[7]

The result of these interdependent forces was a dispersed settlement pattern, which forced a dependence on the automobile. In 1972 over 30 percent of all households owned two or more cars—nearly double the percentage reported ten years prior. Households with no cars declined from 24.3 percent of the population to 20.5 percent during the same period.[8]

The stock of privately and publicly owned vehicles increased at an average annual rate of 4.1 percent, which was equal to the rate of growth in real GNP and far exceeded the rate of growth in population.[9] Meanwhile, the average number of miles driven per vehicle increased from 9,646 to 10,370 from 1963 to 1972.[10] Automobiles in two- and three-or-more car households averaged more miles *per vehicle* annually than automobiles operated by one-car households.[11] Also during this period the average number of miles traveled per gallon dropped from 14.37 to 13.49 for all passenger cars as rising incomes led to increased purchases of larger cars.[12]

---

[a]This theory is presented in William Alonso, *Location and Land Rent* (Cambridge: Harvard University Press, 1964). The empirical support can be found in Richard F. Muth, *Cities and Housing: The Spatial Pattern of Urban Land Use* (Chicago: University of Chicago Press, 1969), p. 324.

Thus, rising incomes and falling relative energy prices, by 1972, had led to a situation in which transportation energy had been substituted for a more compact form of settlement. Not only did this increase the consumption of gasoline at a rapid rate, but it also reduced the responsiveness of the system to changes in these prices. Walking, and to some extent mass transit, were no longer easily substituted for the automobile, since origins and destinations were so dispersed.

The Fuel Composition of Demand

Energy is not a homogeneous commodity. Each source of energy has its own properties and history. As can be seen in Table 2-2 in the last two or three decades there has been a rather dramatic shift away from coal and toward petroleum and natural gas. Just as coal had replaced lumber as a major source of energy so it, in turn, was replaced by oil and natural gas. Coal replaced lumber largely because lumber was growing increasingly scarce and because of the invention of the steam engine.[13] The steam engine increased the economic desirability of coal, not only by lowering its production and transportation costs, but also by providing a new, important market for it. Domestic exploration and mining activity, responded to this technological advance and, by 1920, the United States accounted for 45 percent of the estimated world production of coal.[14]

The story of the emergence of natural gas and petroleum as the dominant forms of energy are intertwined because they frequently are found together in the same geologic formations.[b] The demand for petroleum products, particularly

Table 2-2
Fuel Shares of Total BTU Energy Consumption for Selected Years
(Percentages)

Year	Coal	Petroleum	Natural Gas	Nuclear & Hydro	Total
1950	37.7%	39.8%	18.2%	4.3%	100.0%
1962	21.3	45.0	29.8	3.9	100.0
1972	17.3	45.7	32.1	4.9	100.0

Source: Federal Energy Administration, *Project Independence Report*, Washington: U.S. Government Printing Office, November 1974), appendix pages 9, 13, and 16.

---

[b]The natural gas found with crude oil is known as associated gas. In 1974 associated natural gas accounted for approximately 18 percent of all natural gas produced. See Federal Administration, *Project Independence Report* (Washington: U.S. Government Printing Office, 1974), pp. 93-94.

gasoline, grew sharply in response to the growth of the automobile industry. Once gasoline had become such an important product in a use with which coal could not effectively compete more exploration and refining activities were initiated. This exploration activity resulted in the discovery of large quantities of natural gas.

The displacement of coal from its position as the nation's primary energy source had its roots in the transportation sector, but it soon spread to other types of uses. Coal has a high volume per BTU when compared to oil and natural gas. In addition it is a solid form of energy. These physical characteristics make it more difficult to transport and use. The use of coal involved a more labor-intensive process than the use of other fuels both in terms of adding the fuel to the combustion process and in terms of removing the solid residue that remained after combustion. In comparison, feeding oil and gas to the combustion process automatically was easier and cheaper. In addition, oil and natural gas burned cleaner, obviating the need for additional solid waste disposal expenditures.

As natural gas was discovered, it replaced manufactured gas and some coal in the geographic areas near where it was found. Then, with more natural gas available, and a developing geographically dispersed demand for it, a long-distance gas transmission system was constructed. The technology for these new lines was developed in the 1930s, but it was not until after World War II that the rapid transition from manufactured to natural gas was begun.[15]

The substitution of natural gas and petroleum products for coal, which started in the 1920s, was reinforced in the 1960s by the behavior of coal prices. The wholesale price index for coal increased 72 percent from 1969 through 1972. During this time the wholesale price index for refined petroleum products rose only 9 percent.[16] This larger increase in coal prices has been attributed to the passage of Mine Health and Safety Act of 1969, which increased the costs of mining.[17]

Sources of Supply

The combination of an increasing demand for energy, in general, and an increasing share of that market being captured by oil produced a level of domestic consumption that exceeded domestic production. As a result, imported petroleum products rose from 7 percent of total consumption in 1947 to 30 percent in 1972.[18]

This increasing reliance on imports can be explained, in part, by explicit government policy. In the case of crude oil this policy dates back to the end of World War I. Because, even at that time, there was great fear that domestic reserves would soon be exhausted, the United States government took two major steps to solve this problem. The first was a systematic program to open the oil reserves in other parts of the world to United States use and the second was a program to artificially hold down domestic production rates. The United States State Department actively aided American companies in their attempt to enter Iraq, the Netherlands, East Indies, Mexico, Kuwait, and Venezuela in the period between World War I and World War II.[19] This diplomatic assistance was supplemented by a tax policy that encouraged overseas investment. United States companies were allowed to claim depletion allowance on foreign reserves and, perhaps more importantly, were allowed to deduct royalties paid to foreign governments from their United States tax liability. The intent of these actions was to induce United States energy corporations to increase their foreign investments. It worked. By 1960 about one-third of the book value of all United States foreign investments was accounted for by petroleum.[20]

Simultaneously action was taken on the domestic front to lower domestic production rates.[c] In part this was due to the phycial geology of oil reserves. The level of ultimately recoverable reserves in a given reservoir is a function of the rate of production out of that well. Very high production rates can lower the percentage of oil in the reservoir that could be ultimately recovered. With a single owner this would not be a problem because that owner could adjust his rate of production so as not to exceed the maximum efficient rate. In practice, however, a single reservoir may be tapped by many owners and, without explicit cooperation among them, each owner has an incentive to produce somewhat more than his share of the maximum efficient production rate. Otherwise the conserved oil may be pumped by one's competitors.

The government intervention initiated in response to this problem represented a collaborative effort between the federal and state governments. The state governments took the lead by imposing market demand prorationing on the producers within their states. This device specified both the aggregate amount of production that would be allowed and its allocation among wells. Oklahoma issued the first statewide proration order in 1928.[21] Other producing states soon followed suit.

The federal government assisted the states in their efforts to curb production first by the passage of the National Recovery Administration Petroleum Code, adopted late in 1933, and, subsequently, by the Connally "Hot Oil" Act in 1935. The broad purpose of these acts was to strengthen the conservation program by removing the incentive for states to compete among themselves for a larger share of the market, which, if uncontrolled, could undermine the whole effort, and to protect against price undercutting by foreign producers.

The effect of these laws was not only to limit production, the intended result, but also to keep United States crude prices higher than their competitive level. In addition, it was not clear that the initial physical problem that justified

---

[c]This section relies heavily on Stephen L. McDonald, *Petroleum Conservation in the United States: An Economic Analysis* (Baltimore: Johns Hopkins Press for Resources for the Future, 1971).

conservation, overproduction leading to lower ultimate recovery, was resolved by the regulations because of distortions caused by the specifics of the prorationing system.[22]

During this same period the government also became an active participant in the natural gas market.[d] The milestones in terms of its increasing role were the Natural Gas Act of 1938 and the 1954 Supreme Court decision in *Phillips Petroleum Company* vs. *Wisconsin.* The Natural Gas Act transformed the Federal Power Commission (FPC) into a natural gas regulatory agency, charged with maintaining just prices. In fulfilling the requirements of this act, the FPC limited its price control activities to the pipeline companies. In the 1954 decision the Supreme Court ruled that the Natural Gas Act of 1938 had been misinterpreted by the FPC and that they were required to regulate producer (wellhead) prices as well.

This regulation has resulted in two main sources of inefficiency: On the one hand the artificially low price prevented supplies from increasing as rapidly as they otherwise would have. Stephen G. Breyer and Paul W. MacAvoy have estimated that by 1968 additions to reserves might have been three times greater and immediate production twice as great if there had been no field price regulation.[23] On the other hand the low price served neither to restrain demand nor to allocate the scarce natural gas. As a result shortages occurred and the available supplies had to be administratively rationed. Since the 1960s one of the chief rationing devices has been the refusal to provide natural gas to a broad class of potential new users.[24] To the extent that these users would use the natural gas more productively than the established users, a resource misallocation results; these more productive users are not allowed to bid the available supplies away from the less efficient users.

The increasing reliance on crude oil imports brought about by these policies caused two problems: In the first place it was recognized that foreign control over a crucial input to the United States productive process created a potentially serious natural security problem. Second, since the imported crude was generally cheaper than the domestic crude produced under the prorationing system, these imports posed a potentially serious threat to the domestic oil industry.

As a result, there was somewhat of a reversal of the international policy and controls were placed on oil imports. The first system of controls was imposed by Secretary of the Interior Harold Içkes on September 2, 1933.[e] These were subsequently abandoned in 1935 after the Supreme Court ruled the National Industrial Recovery Act, the legislative basis for import controls, unconstitutional. A system of voluntary import quotas was implemented by President

[d]This section relies heavily on Stephen G. Breyer and Paul W. MacAvoy, *Energy Regulation by the Federal Power Commission* (Washington: Brookings Institution, 1974).

[e]The political maneuvering behind this and related measures are described in detail in George D. Nash, *United States Oil Policy 1890-1964* (Pittsburgh: University of Pittsburgh Press, 1968).

Eisenhower on July 30, 1954; these were replaced by mandatory import quotas on March 10, 1959.[25] While the import controls may well have maintained imports lower than would have been the case without them, it did not prevent the increasing dependence on them. By September 1973 we were importing about 38 percent of our consumption of all petroleum products.[26]

Thus, the stage was set for the 1973 Arab oil embargo. The country was heavily dependent on natural gas and petroleum. Because of structural changes in the economy, stimulated by rising income and falling relative prices, the ability to adjust quickly to sudden changes in the supplies of these fuels had been lost. The demand for energy was growing faster than domestic production, creating an increasing dependence on foreign countries for these products. In the fall of 1973 the imports from many of the Arab countries were embargoed and, subsequently, all imports were subject to a significant price rise. Chapter 3 relates the story of that embargo and how it affected energy policy.

Notes

1. U.S. Department of Commerce, *Business Statistics: The Biennial Supplement to the Survey of Current Business* (Washington: U.S. Government Printing Office, 1973), p. 44.

2. Ibid., p. 46.

3. Bruce Hannon, "Energy Conservation and the Consumer," *Science*, 11 July 1975, p. 96.

4. Ernst R. Berndt and David O. Wood, "Technology, Prices, and the Derived Demand for Energy," *Review of Economics and Statistics* (in press).

5. Washington Center for Metropolitan Studies, "Lifestyle and Energy Surveys," reported in the Ford Foundation Energy Policy Project, *A Time to Choose* (Cambridge: Ballinger, 1974), p. 118.

6. Hannon, "Energy Conservation and the Consumer," p. 98.

7. See J.R. Meyers, J.F. Kain, and M. Wohl, *The Urban Transportation Problem* (Cambridge: Harvard University Press, 1965).

8. Motor Vehicles Manufacturers Association, *1973-74 Automobile Facts and Figures* (Detroit: Motor Vehicles Manufacturers Association, N.D.).

9. Ibid., p. 16.

10. Ibid., p. 44.

11. U.S. Department of Transportation and the Federal Highway Administration, *Nationwide Personal Transportation Study: Annual Miles of Automobile Travel (Report No. 2)* (Washington: U.S. Government Printing Office, 1972), p. 3.

12. U.S. Department of Transportation and the Federal Highway Administration, *Highway Statistics, Summary to 1965* (Washington: U.S. Government Printing Office, 1967), p. 43, and *1972 Highway Statistics* (Washington: U.S. Government Printing Office, 1973), p. 52.

13. Nathan Rosenberg, "Innovative Responses to Material Shortages," *The American Economic Review* LXIII (May 1973): 114.

14. W.N. Peach and James A. Constantin, *Zimmerman's World Resources and Industries*, 3rd ed. (New York: Harper & Row, 1972), Table 28.1, p. 364.

15. Federal Energy Administration, *Project Independence Blueprint Final Task Force Report: An Historical Perspective* (Washington: U.S. Government Printing Office, 1974), p. 25.

16. U.S. Department of Commerce, *Business Statistics: 1973*, p. 46.

17. Richard L. Gordon, "Coal's Role in the Age of Environmental Concern" in Michael S. Macrakis, ed., *Energy: Demand Conservation and Institutional Problems* (Cambridge: MIT Press, 1974), p. 229.

18. U.S. Department of Commerce, *Business Statistics: 1973*, p. 166.

19. Edward H. Shaffer, *The Oil Import Program of the United States: An Evaluation* (New York: Frederick A. Praeger, 1968), p. 13.

20. Gerald M. Brannon, *Energy Taxes and Subsidies* (Cambridge: Ballinger, 1975), p. 92.

21. Stephen L. McDonald, *Petroleum Conservation in the United States: An Economic Analysis* (Baltimore: Johns Hopkins Press for Resources for the Future, 1971), p. 37.

22. Ibid., p. 185.

23. Stephen G. Breyer and Paul W. MacAvoy, *Energy Regulation by the Federal Power Commission* (Washington: Brookings Institution, 1974), p. 83.

24. Richard B. Mancke, *The Failure of U.S. Energy Policy* (New York: Columbia University Press, 1974), p. 115.

25. Shaffer, *Oil Import Program*, pp. 18-22.

26. Federal Energy Administration, *Monthly Energy Review* (November 1974): 8-10.

3

Rearranging the Policy Agenda: The 1973 Oil Embargo

In Chapter 2 we dealt with the factors that led to an increasing importance of energy in the economy, an increasing reliance on petroleum products and natural gas as the primary sources of that energy, and an increasing dependence on foreign suppliers. In the classical economics world of free trade this was a desirable exploitation of the principle of comparative advantage. Nations specialize and trade with each other to gain the products they do not produce. Self-sufficiency is specifically rejected as inefficient and wasteful. By skillful trading the United States could use the inexpensive oil from the Middle East and South America to produce machinery and food, which could be exported to these countries to pay for the oil. Both sets of countries would gain.

The beauty of free trade, however, is marred when the trading relationship is subject to political control. Since this political control is exercised by a foreign power that does not necessarily share the long-run goals and interests of the second country, serious conflicts can result. In the fall of 1973 one such conflict arose. A coalition of Arab oil producers raised crude oil prices dramatically and imposed an embargo on shipments of crude oil from their countries. The purpose of this chapter is to explain briefly the causes and implications of that embargo. Several questions guide the inquiry: How did the participating countries gain enough power to exercise so much control over the world petroleum market? What triggered the embargo? What costs did the embargo impose on the United States? How did the embargo affect long-run energy policy?

Political Background

The conscious policy of encouraging energy investment abroad was facilitated in the early 1900s by the ability of the major United States oil companies, in conjunction with the State Department,[1] to establish concessions in countries with, what were later discovered to be, very large oil reserves. These concessions gave the major oil companies access to large geographic areas of potentially productive oil bearing geologic formations for a large number of years. The companies were granted almost total control over production from these fields and the pricing policies to be followed. In return for these concessions the host governments (or rulers) commonly received fixed production royalties, loans against future royalties and rental payments on the land until royalties began to flow in.[a] Initially the arrangements satisfied both parties. The host countries needed revenue and had neither the financial capital nor the technical expertise to develop their own resources. The benefits to the oil companies of gaining access to these large new fields was obvious.[b]

These arrangements lasted without severe modification until the 1940s. By then pressures for change were growing in the host countries. There were several reasons for this. The fixed royalty payments provided no hedge against inflation. The host countries were selling their resources at a fixed price, while the costs of the goods they were buying were rising. Nationalism was growing stronger in the host countries and this led to a resentment about the degree of control exercised by the foreign oil companies. Finally, since the oil companies had done their job well, substantial reserves were uncovered. It was becoming increasingly clear to the host countries just how valuable these reserves were.

The initial change occurred in Venezuela in 1943. A law passed in that year altered the relationship of the concessionaire to the mineral deposits in the host country. Among other innovations, the law, as amended in 1945, 1947, and 1948, established the principle that profits earned on the crude production were to be shared between the multinational oil companies and the host countries.[2] The inclusion of this principle into concession agreements soon spread to the Middle East countries as well. By 1950, for example, the ARAMCO-Saudi Arabian agreement had been modified to include this provision.[3]

The device used to operationalize this concept was the "posted price." The posted price was administratively determined, although in the earlier years of its use it approximated the realized price. Although the administered price could, and did, change periodically, the use of a posted price, rather than an actual price, was designed to give a certain stability to prices. This stability was valued both by the multinational oil companies, as a device to prevent serious price competition, and by the countries, to protect themselves against price declines.

In 1957 crude oil posted prices were raised. M.A. Adelman attributes this not to scarcity, but rather, to an attempt by the American oil companies to thwart the rapid growth of imports into the United States by American independent refiners.[4] This rise in posted prices was supported by the producing nations because it increased their tax revenue per barrel sold. The industry, however, subsequently changed its mind about the efficacy of higher prices; it discovered that it was not able to translate these higher crude prices into higher product prices because of the introduction of wide-scale discounting.[5]

Therefore, in 1959 the oil companies asserted their power to reduce posted

[a]The ARAMCO concession agreement in Saudi Arabia was fairly typical. See Donald A. Wells, "ARAMCO: The Evolution of an Oil Consession," in Raymond F. Mikesell et al., *Foreign Investment in the Petroleum and Mineral Industries* (Baltimore: Johns Hopkins Press for Resources for the Future, 1971), pp. 216-36.

[b]The American companies were, however, accepting some risk. For example, although the first test drilling occurred in Saudi Arabia in 1934, the first successful wells were not located until 1938. It was not until after World War II that the size of the Saudi Arabian reserve was fully appreciated. See Neil H. Jacoby, *Multinational Oil: A Study in Industrial Dynamics* (New York: MacMillan Co., 1974), p. 36.

prices and this became a major source of irritation to the governments of the producing states. In response, in 1960 the Organization of Petroleum Exporting Countries (OPEC) was formed for the purpose of developing a coordinated, unified position for the producing nations. While the balance of power between the host countries and the oil companies had shifted substantially toward the host countries since the original concession agreements, the oil companies still had the upper hand. They still had the technical expertise and, more importantly, they controlled the distribution and marketing channels. Unlike the host countries, the multinational major oil companies had demonstrated an ability to act in concert.[c]

A necessary condition for the transformation of OPEC into a force to be reckoned with was the dilution of the power of the major integrated oil companies. The dilution of this power was actively sought, particularly by Algeria and Libya, by bringing more independent oil companies into the world petroleum market.[6] As Adelman puts it: "The old internationals were perhaps reluctant rivals; but competition kept breaking or creeping in, and long before the end of the decade they had lost control of the market."[7] Also during this period an Arab-Israeli war caused the closing of the Suez Canal. This increased the attractiveness of Libyan oil for Europe and by 1969 about 30 percent of Europe's needs were coming from that source.[8] Then in May 1970 the Trans-Arabian pipeline was blocked by Syria and the Libyan government began production cutbacks to force an agreement on higher taxes.[9] The result was an increase in the delivered price in Rotterdam of about 50 percent.[10] Since, by this time, demand growth had eliminated any excess production capacity,[11] the oil companies in Libya capitulated and agreed to higher prices. This agreement was soon followed in other countries by similar agreements. The price increases were followed in the next two years by nationalization of certain companies in Iraq and increased participation in others by Libya and Saudi Arabia.[12]

The balance of power had now swung to the host nations, which were by that time the chief oil exporting nations. Oil was vitally important to the consuming nations. In the short run good substitutes were not available. The power of the major international oil companies had been diluted. The oil exporting nations, by 1973, possessed both the ability and the motivation to impose their will upon the world petroleum market.

The Nature and Cost of the 1973 Oil Embargo

There were two main actions taken in October 1973 to assert this power.[d] On

---

[c]In the Achnaccary Agreement of 1928 Shell, B.P., and Exxon reportedly had entered into an agreement to share markets outside the United States and to coordinate facilities to stabilize prices. See Jacoby, *Multinational Oil*, p. 30.

[d]This section is based upon Federal Energy Administration, Office of International Energy Affairs, *U.S. Oil Companies and the Arab Oil Embargo: The International Allocation of Constricted Supplies* (Report prepared for the use of the Subcommittee on Multinational Corporations of the Senate Committee on Foreign Relations, 94th Cong., 1st sess., January 27, 1975, appendix I).

October 16 the decision was taken by the Persian Gulf members of OPEC to raise prices unilaterally by 70 percent. On the following day this action was ratified by OPEC as a whole. In addition, the Arab states undertook a series of production cutbacks and attempted to target the shortfall by a complete embargo of all exports to the United States, Canada, the Bahamas, Trinidad, the Netherlands, Antilles, Puerto Rico, Guam, and the Netherlands. Other nations were eventually placed in one of three categories: (1) most favored, (2) preferential, or (3) neutral. The "most favored" nations received oil enough to meet their current demands. The "preferential" nations received approximately their September 1973 levels while the "neutral" nations were allocated the remaining oil, prorated on their September 1973 import levels. The Arab nations made clear that a nation's position on the Arab-Israeli dispute would influence its classification within these categories. In addition, following a November summit meeting in Algiers, the Saudi Arabian and Algerian oil ministers toured Europe, explicitly linking the restoration of production to a pullback by Israel from the occupied territories. On December 22, 1973 posted prices were again raised, this time by 130 percent. The embargo against the United States was lifted by most Arab states, as abruptly as it had begun, on March 18, 1974. The price hikes remained in force.

This series of events affected the American economy in four independent ways: (1) it reduced the supplies of energy available to American industry and consumers; (2) it simultaneously increased the cost of energy, which, because of incomplete substitutability, reduced real consumption levels; (3) it caused the transfer of a large amount of income from the United States economy to the OPEC nations; and (4) it created an uncertain environment, which affected expectations and behavior. The reduction in supplies started slowly but reached a high in excess of 3 million barrels a day during February and March of 1974. This represented a shortage of about 17 percent of the expected domestic consumption of petroleum during that period.[13] The accompanying price rises, which affected non-Arab, as well as Arab, oil, were dramatic. The landed price of crude oil rose from approximately $3.50 a barrel in September 1973 to over $13.00 a barrel by the following May.[14] Although the prices received by the oil companies for domestically produced oil were controlled and nonmarket procedures used to allocate the shortages, the increases in costs due to the rise in the prices of imports were allowed to be passed through to end users. The result was that the annual inflation rate, as measured by the Consumer Price Index, was 12.1 percent in the first quarter of 1974 instead of the previously forecasted 5.1 percent.[15] The unemployment rate went from 4.6 percent for all civilian workers in October of 1973 to 5.2 percent by May 1974.[16]

The costs that OPEC was able to impose on the United States economy are important because they provide the means to empiricize and operationalize the concept of import vulnerability. They provide a benchmark for forecasting the costs of future embargoes. These costs in turn provide a basis for judging how much the United States should be willing to spend in reducing import vulnerability. Unfortunately, the estimation of this cost, even retrospectively, is as difficult as it is important.

The main difficulty in measuring the cost of the embargo lies in accurately characterizing what would have happened in the absence of the embargo. To the extent that the economy was on the brink of recession without the embargo the estimated cost of the embargo (the difference between the growth path if the embargo had not happened and the actual growth path) would be smaller than if the economy was on a strong growth path. Most observers believe that the former is a more accurate characterization of the period than the latter. Some estimates of the cost of an embargo derived from large-scale macroeconometric models are provided below in Table 3-1.

The anticipation of these costs and the desire to minimize them precipitated rapid government action in the United States. On November 17, 1973 President Nixon signed the Mandatory Fuel Allocation Act, which directed him to institute a mandatory allocation program for crude oil and petroleum products. The administration program was designed to protect certain priority categories of fuel use (e.g., industrial uses and residential heating fuels) from scarcity, insofar as possible, by channeling most of the shortfall toward gasoline.

Table 3-1
Estimates of the Cost of the 1973 Arab Oil Embargo in Terms of Reduced Real GNP

(Billions of 1973 Dollars)

Source of Estimate	Quarter					
	73:4	74:1	74:2	74:3	74:4	75:1
Data Resources, Inc.[a]	$5.9	$32.1	$30.1	$26.4	$21.8	$24.2
Department of Commerce[b]	3.2	16.1	15.0	15.7	22.4	23.1
Enzler-Pierce[c]	4.5	10.6	17.6	22.7	27.5	32.2
Perry-FRB/MIT[d]	5.4	16.0	24.1	29.6	35.6	37.0
Perry-Michigan Model[e]	3.9	9.1	15.1	21.5	33.2	39.8

Note: Computed as the difference between an embargo and a control forecast. The Perry estimates were presented in 1973 dollars. All other estimates were converted from 1958 dollars by multiplying by 1.543, the 1973 GNP deflator.

[a]Reported in Federal Energy Administration, *Project Independence Report* (Washington: U.S. Government Printing Office, 1974), p. 292.

[b]Ibid.

[c]James L. Pierce and Jared J. Enzler, "The Effects of External Inflationary Shocks," *Brookings Papers on Economic Activity, 1974:1*, p. 48.

[d]George L. Perry, "The Petroleum Crisis and the U.S. Economy," to appear in Edward R. Fried and Charles L. Schultze, eds., *Higher Oil Prices and the World Economy* (Washington: Brookings Institution, forthcoming), table 6.

[e]Ibid., table 7.

The mechanism chosen to regulate gasoline consumption was to ration the available supplies to wholesalers and then to place some, but not complete, restrictions on how the wholesalers dispensed the gasoline. The wholesalers were required to supply, from their allotment, 100 percent of base period use (usually the amount supplied to that user in the corresponding month one year earlier) to certain designated key businesses. What was left could be rationed by the wholesaler in a variety of ways, but price rises were strictly limited.

As a result gasoline shortages did appear, although their severity varied markedly from region to region. There were various rationing devices employed by the wholesalers including Sunday closings, priority for established customers, maximum and minimum limits on purchases, etc. The nonuniformity of the practices caused a great deal of uncertainty among drivers and served to restrict vacation travel a good bit.

Although the government was not very successful in shifting refinery production from gasoline to the other, higher priority petroleum products,[17] gasoline did register the largest shortages. The employment effects of the embargo reflect this orientation. The desirability of gasoline as the fuel to take the brunt of the shortage was based in part on the assumption that the employment effects of a shortage of gasoline would be smaller than for other fuels.[e] While this rationale is probably correct, the gasoline induced employment effects were discernible.

Most of the unemployment caused by the embargo and the allocation program hit blue-collar workers.[18] The shortage of gasoline caused an estimated 10.3 percent decline in employment by gasoline service stations, and the combination of a shortage with higher gasoline prices (resulting from the pass-through of higher imported oil prices) were chiefly responsible for a 14.8 percent decline in motor vehicle and equipment manufacturing, a 6.4 percent decline in motor vehicle retailing employment and a 3.0 percent decline in employment in the hotel industry.[19] Sales and employment in the automobile industry and its related industries, were, in retrospect, quite sensitive to the gasoline shortage and price rises. The sales of new automobiles during the period reflected a much greater interest in smaller, gas-efficient automobiles than historically had been the case.[20]

The Embargo in Historical Perspective

The 1973 embargo was an important milestone in several respects. In the first place it provided a recent historical experience that could be used to assess the importance of the import vulnerability problem. It provided a concrete example

---

[e]For an examination of the effects of shortages of other fuels on heavy energy-using industries, see Federal Energy Administration, Office of Economic Impact, *Short-Term Microeconomic Impact of the Oil Embargo, October 1973-March 1974* (Washington: U.S. Government Printing Office, n.d.).

of what, prior to that time, had been considered a somewhat remote possibility. It helped shape perspectives on what to expect in the future. It also fundamentally altered the nature of the problem. Prior to the embargo the import vulnerability problem had been thought of in terms of cheap imports increasingly replacing domestic sources, leaving the country heavily reliant on foreign sources. After the very heavy price increases accompanying the embargo, the problem became more one of substituting cheaper domestic sources for the more expensive foreign ones. Finally, the embargo gave a good deal of visibility to the energy problem and guaranteed that the problem would get political attention, if not political action.

The embargo made clear the fact that the stability of import prices and the security of foreign supplies were not matters to be treated lightly. The adverse effects were potent and politically volatile. It also provided a benchmark for the assessment of policy alternatives. The higher the costs that could be imposed on the United States by the OPEC nations and the more likely these costs would again be incurred, the more willing the nation should be to incur other costs to protect against this vulnerability.

The effects of the OPEC actions, taken in the fall of 1973, are still being felt as this book is being written in the fall of 1975. This gives rise to the issue of whether future OPEC uses of the oil weapon could lead to similar persistent effects on the economy. If they could, then the United States would be justified in making very large expenditures to reduce import vulnerability. In fact, there is reason to believe that, while the action of 1973 may well have caused a very persistent displacement of the United States economy from its long-term growth path, and, therefore, inflicted considerable damage, it is probably wrong to attribute this kind of vulnerability to the future. The reason is a simple one. One characteristic of the last embargo, the one apparently leading to the persistent nature of the costs, is unlikely to be repeated in future embargoes. The prices paid for imported petroleum were maintained at the embargo levels long after the embargo ended. Because the replacement of those imports from domestic sources occurs with such a long time lag and because the demand elasticity is so low, the large outflow of dollars to the OPEC nations, which began during the embargo, has continued. This outflow of dollars, to the extent that it is not compensated for by an increase in exports, lowers the demand for domestic products and precipitates multiplier effects on production and employment.

Future embargoes are likely to be accompanied by price increases for imported petroleum products, but these increases are not likely to persist much beyond the embargo period. The basis for this assessment is that while short-run demand and supply elasticities are very low, the long-run elasticities are not. This implies that by pushing prices much higher than they are now in real terms[f] for a

_____

[f]Note that the maintenance of the 1973 price levels in real terms in an inflationary economy means rising nominal prices. Thus, some price rises by OPEC can be expected even in the absence of embargoes, but these are likely to be much smaller than the 1973 price rises, and, therefore, less harmful to the economy.

long period, the Arab nations would effectively price themselves out of the market. This would deprive them of both the revenues and the political leverage that dependence on their imports brings. It is unlikely, therefore, that the conditions of the last embargo will be repeated.

The fact that the embargo altered the nature of the basic problem was not fully appreciated for some time and, to some extent, still is not. The studies prior to the embargo had typically extrapolated historical trends of consumption and production with the result that United States dependence on foreign oil would increase dramatically over the next decade. This notion of the problem was ingrained in Washington thinking and perspectives change slowly. After the embargo, when foreign oil prices were in the neighborhood of $11 a barrel, most domestic oil was fully competitive with imported oil. The economic viability of even the more exotic fuels became a strong possibility if these high prices would persist. In short, the oil price rises could be expected to unleash to a considerable extent, the domestic price system. The expected increases in production and reductions in consumption in response to these higher prices, if they are maintained, would go a long way toward solving the problem of import vulnerability.

The final impact of the embargo was to give the energy problem public visibility. The desirability of creating a national energy policy was not an idea that originated in the Nixon administration. In 1952 the prestigious Paley Commission wrote:

The Commission therefore concludes that the most important step for Government to take at this time toward developing a comprehensive energy policy is to achieve through a single agency, a comprehensive and continuing review of the long-term energy outlook and an appraisal of the adequacy of public and private policies and programs for coping with the problems that such a review may reveal.[21]

Twenty-two years later the *Project Independence Report* was issued. Two conditions seemed to jolt the government into action. On the one hand, large numbers of people were affected by the embargo and this focused their attention on energy policy. Surveys taken during the period indicate that during February 1974 nearly 70 percent of a national sample indicated that their life had changed for the worse as a result of the embargo.[22] In addition, this mobilized public opinion placed the blame for the energy situation squarely on the shoulders of the United States government. During February the same national sample was asked, "Which group is most responsible for the current energy shortage?" The government in Washington received the greatest percentage of responses (45 percent), oil and gas companies were named next most often (32 percent), and the Arabs were a distant third (10 percent).[23]

The government was not insensitive to this developing political climate and moved rather rapidly in the face of it. On November 7, 1973 President Nixon delivered a nationwide radio and television address urging domestic energy conservation and proposing Project Independence, which was to be a massive effort, comparable to the Manhattan Project, designed to achieve energy self-sufficiency by the 1980s. Chapters 4, 5, and 6 provide an examination of the analytical response to this challenge.

Notes

1. Harold F. Williamson et al., *The American Petroleum Industry, 1899-1959: The Age of Energy* (Evanston, Illinois: Northwestern University Press, 1963), pp. 517-19.

2. Fuad Rauhani, *A History of O.P.E.C.* (New York: Praeger Publishers, 1971), pp. 45-46.

3. Donald A. Wells, "ARAMCO, The Evolution of an Oil Concession," in Raymond F. Mikesell et al., *Foreign Investment in the Petroleum and Mineral Industries* (Baltimore: Johns Hopkins Press for Resources for the Future, 1971), p. 220.

4. M.A. Adelman, *The World Petroleum Market* (Baltimore: John Hopkins Press for Resources for the Future, 1972), p. 161.

5. Ibid., p. 161.

6. Federal Energy Administration, *Project Independence: An Historical Perspective* (Washington: U.S. Government Printing Office, 1974), p. 3.

7. Adelman, *The World Petroleum Market*, p. 254.

8. Federal Energy Administration, *Project Independence: An Historical Perspective*, p. 4.

9. Adelman, *The World Petroleum Market*, p. 25.

10. Ibid.

11. James W. McKie, "The Political Economy of World Petroleum," *The American Economic Review* LXIV (May 1974): 51.

12. Joseph A. Yager and Eleanor B. Steinberg, *Energy and U.S. Foreign Policy* (Cambridge: Ballinger, 1974), pp. 16-17.

13. Federal Energy Administration, *Project Independence Report* (Washington: U.S. Government Printing Office, 1974), appendix, p. 284.

14. Federal Energy Administration, *Monthly Energy Review* (December 1974): 41.

15. Federal Energy Administration, *Project Independence Report*, appendix X, p. 295.

16. U.S. Department of Commerce, *Survey of Current Business* (October 1974): S-13.

17. George L. Perry. "The Petroleum Crisis and the U.S. Economy," in Edward R. Fried and Charles L. Schultze, eds., *Higher Oil Prices and the World Economy* (Washington: Brookings Institution, forthcoming).

18. Paul O. Flaim, "Employment and Unemployment During the First Half of 1974," *Monthly Labor Review* XCVII (August 1974): 6.

19. John F. Early, "Effect of the Energy Crisis on Employment," *Monthly Labor Review* XCVII (August 1974): 10.

20. Federal Energy Administration, *Project Independence Report*, appendix, p. 302.

21. U.S., President's Material Policy Commission, *Resources for Freedom: A Report to the President*, vol. 1 (Washington: U.S. Government Printing Office, 1952), p. 30.

22. National Opinion Research Center, *The Impact of the 1973-74 Oil Embargo on the American Household* (Chicago: National Opinion Research Center, 1974), p. 189.

23. Ibid., p. 175.

Part II:
The Analytical Response

4 Conceptualizing the Energy Problem

In examining the responsiveness of political institutions we start first with an examination of the problem of providing policy makers with useful and timely information on the consequences of taking proposed policy actions. In an area such as energy policy where many interdependent policies are likely to be pursued it is impossible for individual policy makers to comprehend the likely overall effects of a proposed program without the assistance of some internally consistent mechanism for relating ends to means.

As argued in Chapter 2 it is by no means automatic that an analytic effort of this magnitude can be mounted in time to be of assistance or, even if it can, whether it would be used. In Part II the analytical response to the need for both a bank of substantive information and a flexible modeling system to relate ends to means is described.

Any analytical modeling effort is shaped by the purposes for which the model will be used. An all-purpose model, even if one existed, would certainly be overly general for most particular problems and, hence, of limited value for those applications. Models are rarely perfectly malleable.

Since this is true, the analyst's conceptualization of the problem is important as a guide for specifying the configuration of the modeling system to be used. Since no model can do all things, it serves as a filter for separating the important from the less important dimensions. It allows the anticipation of particular information needs on the part of policy makers.

In this chapter the conceptualization that guided the Project Independence analysis is discussed. This discussion serves a two-fold purpose within the content of the book as a whole: it provides the reader with an overall introduction to the various dimensions of the energy problem and it lays the groundwork for discussing the various choices made in setting up the modeling system. The discussion opens by specifying the various dimensions of the energy situation that suggested the need for long-range planning. This is followed by a brief discussion of four basic strategies that could be pursued in meeting these problems. Finally the key variables used to characterize the behavior of the energy system are enumerated and the role of the government in guiding this system toward the desired goals is described.

Central Policy Dimensions

As the modeling effort took shape several policy dimensions emerged as being of primary importance. These dimensions indicate the complex relationships among

35

foreign policy, energy policy, and economic policy, which must be dealt with by any serious study of energy policy options. They can be described as the import vulnerability problem, the balance of payments problem, and the foreign policy problem.

The import vulnerability problem was foreshadowed in Chapter 3. As the Organization of Petroleum Exporting Countries (OPEC) nations gained effective control over the world petroleum market, they gained control over a key input in the United States production process. This meant, for the United States, that the price and availability of foreign oil was subject to potentially large fluctuations. Aside from making planning more difficult, both for domestic producers and users of energy, it made achieving the goal of economic stabilization more difficult, since the economy was subject to potentially large, externally generated shocks.

The balance of payments problem stemmed from the large price increases instituted by OPEC in 1973.[a] These increases were accomplished so rapidly and were so large in magnitude that the consuming nations were expected to be plunged into a stubbornly persistent current account deficit in their collective balance of payments.[b]

This development caused two potential problems for oil consuming nations. The first and most immediate problem would be to insure that the deficits could be financed by each of the deficit nations. Since the foreign exchange to pay for the increased cost of imports could not solely come from exports, it had to come from some other source. Only for the major countries are their domestic currencies considered an acceptable international form of payment. Foreign exchange could be borrowed by the deficit nations as long as their credit worthiness had not been undermined by their current account deficit. Failure to secure adequate financing would necessitate sharply curtailing imports, which could have serious effects on the local economy and eventually on the world economy.

But the ability to finance the deficit is only part of the problem. Sooner or later these financial resources would be cashed in by OPEC for real resources. The effects of these real resource flows on the oil consuming nations would depend on how the oil exporting nations chose to handle their current account surplus.

Since the surplus nations would, most likely, not be able to translate all financial assets into real assets in the first few years, some would be held, for a while at least, in financial assets. Aside from the potential liquidity problem

[a]This section on the balance of payments problem has benefited from my reading of Dr. Jan Tumlir, "Oil Payments and Oil Debt in the World Economy," *Lloyds Bank Review* (July 1974): 1-14.

[b]A current account deficit occurs when the value of imported goods and services for a nation or group of nations exceeds the value of exported goods and services. This deficit is mirrored by a current account surplus for the rest of the world with whom the deficit nations trade. In this instance the surplus was mainly held by the oil exporting nations.

faced by selected deficit nations this creates two additional kinds of problems for the consuming nations as a group. The concentration of such a huge amount of liquid assets in a limited number of money markets could pose a significant threat to the stability of the world's financial system. The smooth operation of that system depends on a continuing balance between the outflows for loans and the inflows from deposits. If a large amount of short-term funds were suddenly withdrawn from the system, this balance would be upset.

The second problem resulting from the accumulation of wealth in financial assets is the deflationary affect it has on the deficit nations and, potentially, through feedbacks, on the world economy. These huge payments, unless compensated for by an expansionary fiscal and monetary policy, take money that would have been spent domestically out of the domestic economy. The drop in domestic spending could cause a drop in production, leading in turn to an increase in unemployment. This deflationary situation could be transmitted to the rest of the world economy through a reduction in world trade activity.

Instead of, or, in addition to, choosing to invest in short-term financial assets, OPEC could choose to invest the surplus buying ownership claims on corporations in foreign countries. This would give them a claim on productive assets in these other countries and a stream of income from their investments. The potential disadvantage of this choice, from the point of view of the consuming nations, was that foreign control of key domestic industries could give the OPEC nations yet another lever on the economies of the consuming nations.

The third choice OPEC could make would be to use the surplus to increase the flow of imports into their countries. For the countries that would supply the exports (most likely the industrialized nations) this would solve the deflation problem because aggregate demand would be high, production would continue unabated, and employment could remain high. However, the amount of goods and services *domestically consumed* by that exporting nation would fall because a larger proportion of domestic production would be going to OPEC rather than to its own consumers.

In summary the balance of payments problem was anticipated at the time to be potentially quite serious on a worldwide basis, depending on how OPEC dispensed its surplus funds and how the individual countries chose to cope with their deficits.

The third major dimension of the energy situation was its impact on American foreign policy. This dimension is highly related to the first two. The dependence upon OPEC for a critical input made the United States politically vulnerable on several fronts. The threat of other embargoes was now a credible threat and, therefore, could be used to gain leverage in shifting the American position particularly with respect to the Arab-Israeli conflict in the Middle East. Second, the large dollar flows to the OPEC nations gave them an enormous financial resource base. This could be used to gain political leverage in several ways. The money could be spent on arms, perhaps even the technology to produce nuclear weapons, which, if uncontrolled, could produce a fundamental shift in the balance of power, not only in the Middle East, but in the world. The money could be used to buy major international corporations, which, perhaps, could gain control of other markets as well as petroleum. The strategic placement of the funds in short-term financial assets could be used as a basis for blackmail due to their importance in the world financial markets. Sudden shifts of large quantities of short-term deposits could put the affected banks in a precarious financial position. The foreign policy vision was that the very independence of United States foreign policy and the ability to coax the international forces to act in ways congenial to the long-term interests of the United States were in jeopardy.

Strategy Options

In developing a policy evaluation system one has to anticipate the uses to which it will be put. There have to be specific linkages between the policies considered and the structure of the model. If the structure is not compatible with those linkages then the model is ineffectual.

It was anticipated that the United States would have four main energy strategies that it could use to resolve the three dimensions of the energy problem plus a host of complementary policies (e.g., monetary and fiscal policy, etc.). These can be broadly classified as: (1) import substitution, (2) demand management, (3) emergency programs, and (4) foreign policy initiatives. The *import substitution* strategy attempts a resolution of the basic energy problems by reducing imports. This import reduction is accomplished by replacing imported fuel sources with domestically produced fuel sources. These can either be identical in form to the imported fuels (e.g., petroleum) or a different form of energy (e.g., coal). There are several costs associated with this strategy. The increased extraction, refinement, and transportation of domestic energy resources usually imposes higher costs in the form of environmental damage. Although the damage can frequently be controlled, it is controlled only at additional cost. Therefore, whether or not the damage is controlled, costs are incurred. Increased domestic production also leads to a faster rate of domestic resource depletion, which, if serious enough, could substitute future vulnerability for present vulnerability.

The *demand management* strategy also attacks the energy problem by reducing imports, but it accomplishes this reduction in a somewhat different manner. Conservation reduces imports by reducing demand. This can be accomplished by increasing the technical efficiency of the various parts of the energy cycle (hence requiring less energy input per unit output) or by simply consuming less energy (e.g., lowering the thermostat). Demand management has its associated costs as well. Increasing the technical efficiency of energy processes may require large outlays of capital. Reducing the consumption of energy has its own cost, which is reflected in a reduced productivity for industries or a lower level of well-being for consumers.

The third major strategy, *emergency programs*, takes a rather different tack in confronting the problem of import vulnerability. It seeks to reduce vulnerability, not by reducing imports, the approach taken by the two previous strategies, but rather by insulating the economy from the adverse effects of an embargo, should one occur. Because of this difference, unlike the first two strategies, which simultaneously reduce the balance of payments problem and the import vulnerability problem by reducing imports, the emergency programs strategy makes no contribution to reducing the normal dollar outflow for oil nor does it provide protection against permanent foreign price increases.[c] Standby rationing plans to go into effect upon the initiation of an embargo and stockpiles of critical imported materials are examples of programs that also fall under this rubric. Standby rationing plans could reduce the adverse effects caused by a future embargo by injecting more certainty into a highly unstable environment. This could eliminate much of the perverse behavior that seems to occur during embargoes (e.g., hoarding), which tends to exacerbate the damage. These plans could also serve to channel resources toward employment maximizing uses. Stockpiles provide contingency sources of energy, which can replace the embargoed foreign sources in case an interruption in imports occurs. These programs also have their costs. Standby rationing plans entail large administrative costs. Most reasonable rationing programs depend on transferable coupons. These coupons have to be printed and stored and distribution channels have to be established on short notice. If the standby plan is actually implemented, a large bureaucracy will be needed to control the system. Stockpiles require the accumulation of large reserves, which, if no embargo occurs, may never be used.

Foreign policy initiatives can also complement or replace the above strategies. Attempts could be made to orient the import sources away from embargo prone nations toward more secure suppliers. Negotiations could be conducted to reduce the hostility that might lead to an embargo or to politically inspired price increases. The threat of an embargo could be countered by the threat of armed intervention or by retaliatory embargoes on our exports to the nations participating in the embargo.[d] A cooperative effort among consumer nations to provide a countervailing power to the producer nations could serve as the basis for negotiating changes in the world price of oil.

The Role of the Government

Since the modeling system was to be developed to assist in government planning, it had to be based on some notion of the role that the government would play.

[c]In fact, to the extent that foreign oil is used to build a strategic reserve, an emergency programs strategy might exacerbate the balance of payments problem.

[d]The listing of this alternative does not represent an endorsement of it.

There is a presumption in economics that the market system handles resource allocation in an efficient manner unless proven otherwise. The burden of proof therefore, by tradition, has been cast on those claiming that an affirmative role for government is both necessary and judicious.

The case for a positive role for the public sector in shaping the future course of energy in this country is based on four main contentions: (1) there are demonstrable cases where markets inherently fail to provide efficient resource allocations; (2) previous government policy seriously limits the effectiveness of the market in some areas and, hence, the public sector has to either remove or modify those impediments or seek other means to redress the misallocations; (3) there are multiple objectives for energy policy (e.g., foreign policy) and for most of these the public sector is the only legitimate arbiter; and (4) the government is the owner of a large percentage of the remaining energy resources and must decide if and when to produce them. We examine the support for each of these contentions briefly.

The first and most obvious source of market failure occurs in the way environmental costs are handled by the market.[e] Because these costs (e.g., air pollution, water pollution, solid waste, etc.) are most often borne by someone other than the agent causing them, that agent will tend to cause more pollution than is efficient from a larger perspective. This implies that prior to the embargo the observed price of energy products may well have been inefficiently low, since all environmental costs were not correctly internalized.

. The second source of market failure, the one that was used to justify the market prorationing program described in Chapter 2, derives from the fact that oil and gas deposits are frequently found in geologic formations that cover large subsurface areas.[1] Unless the rights to extract from these subsurface areas are under the control of a single owner, conservation incentives are distorted. This is known in the economic literature as the common property resource problem. When a single owner controls the extraction, he will balance current and future production flows to maximize the present value of future profits. Although he is not constrained to limit his production for any physical reason, economic incentives provide a rationale for exercising some restraint. By increasing his rate of production beyond that which maximizes the present value of future profits he would have less oil for sale later when the supplies become scarce and prices high. Compare this situation, however, with a case in which the same source of oil is being tapped by several producers. If one producer decides to conserve for later sale, he has no guarantee that his source will not be drained by others. Thus, by conserving, he could actually forego both current and future income. This kind of incentive structure makes conservation unattractive to the individual producers while it may remain quite desirable to the nation as a whole.

---

[e]One of several good available expositions of this point can be found in A. Myrick Freeman, III, Robert H. Haveman, and Allen V. Kneese, *The Economics of Environmental Policy* (New York: John Wiley and Sons, 1973), pp. 64-79.

The third source of market failure stems from the inadequacy or nonexistence of futures markets in petroleum or natural gas.[f] This is, of course, a crucial deficiency when the markets in question are subject to a great deal of uncertainty. One example of this point lies in examining the question of whether domestic oil producers adequately internalize the risk of an interruption in the supply of foreign oil. There are reasons to believe they do not. For example, one reasonable response to this risk by the producers would be to keep large stockpiles on hand to be used in the event of an embargo. Yet, the oil companies are reluctant to increase their stockpiles to an efficient level because, during embargoes, the nation would not likely allow the price of the stockpiled oil to rise to its market clearing level. Therefore, because of government controls, which remove the profit from stockpiles, it is unlikely that the private sector would provide the efficient stockpile size.

Other misallocations may result from previous government intervention. In Chapter 2 the various forms of regulation over crude oil and natural gas were discussed. The decision to build an energy facility requires compliance with the National Environmental Policy Act. This compliance involves filing an acceptable environmental impact on the project. Nuclear power plants were regulated by the Atomic Energy Commission (now the Nuclear Regulatory Commission). Electricity prices are regulated by state commissions. And so on. However desirable these regulations are on other grounds, they do eliminate, or substantially reduce, the automatic responsiveness of the market system to changing conditions. For this reason the public sector has to reassess continually the value of these regulations in the light of these changing conditions.

The government also has a role to play because energy policy is designed to achieve multiple objectives. Energy policy, as discussed above, is heavily interdependent with foreign policy and economic stabilization. These are goals reserved for the federal government. In addition, as a representative democracy, the federal government is charged with the responsibility of insuring that the distribution of the costs and benefits among different parts of the country and among different socioeconomic groups is just.

Finally, the federal government is drawn into the fray by virtue of the fact that it owns a major share of the resources being considered for use. The major sources of new oil and natural gas are expected to come from offshore areas, which have judicially been determined to be federal properties. Most of the coal reserves lying under the western plains are also owned by the federal government.[2] Since a major concern with using these resources is their environmental impact, a concern not correctly handled by the market system, the government must make the ultimate determination of whether to use these reserves and, if so, when.

The government, therefore, is an inevitable participant in this process. To

[f]This point is discussed somewhat more fully in William D. Nordhaus, "Markets and Appropriable Resources," in Michael S. Macrakis, ed., *Energy: Demand, Conservation and Institutional Problems* (Cambridge: MIT Press, 1974), pp. 16-20.

make informed decisions it must have information on the consequences of its actions. This requires the ability to portray what the energy situation would be like under a variety of different policy choices.

Key Variables

The capability to provide this kind of information presumes a knowledge of the key determining variables and their relationship to the various dimensions of the energy problem. In modeling these relationships one has to decide which variables are to be exogenous and which endogenous.[g] Frequently, as was the case with Project Independence, the conceptual classification of variables into these categories cannot endure the practical realities of modeling; endogenous variables become exogenous, not because the feedback is weak or unimportant, but because the state of the art does not permit the specification of all these relationships with any reasonable degree of confidence.

In the initial conceptualization the prime mover exogenous variables were considered to be the world price for oil, and the various policies undertaken by the United States government. It was clear that the nature and severity of the energy problem depended heavily on the world and the domestic petroleum markets. The decision to make the world price an exogenous variable obviously concentrates the analytic effort on the domestic, rather than the world, market. This choice was made for two reasons: (1) modeling the world market would clearly be a much more difficult venture and probably could not be completed in time to be of much use in the Project Independence Report, which had a November 1974 deadline; and (2) since the world market was dominated by a mixture of political and economic considerations, the causal relationships were much less clear. Therefore, while an analytical effort to model the world economy was initiated, the Project Independence analysis did not depend on it.

The second class of important exogenous variables contained all those variables that could be expected to experience constraints on their behavior. Several potential constraints had emerged in previous studies that could well alter the time path of the system in its pursuit of a long-run equilibrium, if not alter the nature of the equilibrium itself. These constraints fall into three general classes: (1) transitory constraints, which may temporarily restrict the implementation of a strategy; (2) constraints that persist unless the government intervenes; and (3) constraints that are natural and immutable. The first class contains, for example, labor constraints. There is little doubt that high wages in shortage skills would lead eventually to more labor in those areas, but human

---

[g]Endogenous variables are those variables predicted by the model. Exogenous variables, by way of contrast, are taken as given by the model. The exogenous variables then serve as input data and the values of the endogenous variables are estimated conditional on the assumed values of the exogenous variables.

capital is not instantaneously fungible. The second class of constraints contains, for example, transportation and environmental constraints. The market cannot automatically solve these problems because the government exercises a large degree of control over the decisions. Transportation systems are heavily regulated. The building of new transportation systems generally requires at least tacit government approval and usually active participation. Environmental constraints are the direct result of government policy. The air and water quality standards passed by Congress are the most obvious examples. Changes in these constraints in either direction require explicit government action. The final class of constraints are represented by such facts of life as limits on technological efficiency and resource endowments available at particular prices.

The final set of variables, the various energy and economic variables, were to be jointly determined. A limited list would include economic growth, energy prices and quantities by fuel, energy imports and domestic production, inflation, unemployment, interest rates, the balance of payments, etc. The ability to carry out this particular portion of the conceptualization was limited both by the amount of time available for the analysis and by the modeling state of the art.

The limited time available for analysis dictated that the central effort would be focused on capturing the impacts of a limited number of major scenarios. These would at least bracket the important options and give a quantitative assessment of the impacts resulting from them. There would be four major strategies investigated for each of two possible world prices for crude ($7 and $11 in 1973 dollars) and then the outcomes of one of the strategies (the laissez faire or business as usual strategy) would also be estimated for a $4 and a $15 world price for crude oil. The four strategies investigated were: (1) business as usual (BAU), (2) accelerated development (AD), (3) conservation (C) and (4) combined accelerated development and conservation (ADC).

The BAU strategy was supposed to represent what would happen if no new policy initiatives were taken other than removing the existing price controls for domestic crude oil and implementing the deregulation of natural gas. It was a low profile strategy emphasizing only the removal of price system constraints.

The accelerated development strategy represented an attempt to capture the maximum reasonable expansion of domestic energy supplies which could be expected for different world prices for crude oil in the three target years. Examples of the kinds of options considered in this strategy include accelerating Outer Continental Shelf exploration for, and production of, crude oil and natural gas, streamlining siting and licensing procedures for nuclear power plants to reduce lead times, financial incentives for synthetic fuels, financial incentives for solar energy, and additional leasing of coal and oil shale lands.[h]

The conservation strategy represented an attempt to combine a large array of technically and economically feasible policies to reduce energy use.

[h]The exact assumptions can be found in Federal Energy Administration, *Project Independence Report*, p. 64-65.

The conservation strategy included such measures as a 20 mpg standard for automobiles, investment tax credits for energy saving investments (e.g., insulating homes and offices), thermal efficiency and lighting standards and research and development programs designed to increase the technical efficiency of the use of energy in industrial processes.[i]

The final strategy, the combined program of conservation and accelerated development, was the union of the policies embodied in the two strategies described immediately above. This strategy represented the maximum feasible assault on import reduction.

Notes

1. M.A. Adelman, "Efficiency of Resource Use in Crude Petroleum," *Southern Economic Journal* V (1964-65): 103.

2. Federal Energy Administration, *Project Independence Report* (Washington: U.S. Government Printing Office, 1973), p. 103.

[i]The complete program is spelled out in some detail in Federal Energy Administration, *Project Independence Report*, pp. 160-174, and *Conservation Task Force Report*, vols. 1 and 2 (Washington: Government Printing Office, 1974).

5

The Project Independence Report

The analysis that was a part of Project Independence sought to provide two kinds of products to those in a policy-making capacity. The first was to be a comprehensive and comprehensible body of knowledge on the severity of the various dimensions of the energy problem and the effectiveness of alternative strategies in resolving them. The substantive content of that body of knowledge is the subject of this chapter. The second product, the subject of Chapter 6, was to be a responsive and flexible modeling system, which could trace out the implications of various scenarios of interest to policy makers. It represented, in short, an attempt to circumvent the normal problems with incorporating planning into the decision-making process by providing the principals with access to a timely and reasonably flexible modeling system.

Project Independence was conceived in an atmosphere of uncertainty. The Arab oil embargo and the accompanying rise in the world prices for crude oil triggered a large number of concerns about the future, as described in Chapter 4. In the face of these concerns there was a clear need to attempt to separate fact from fiction and the likely from the unlikely.

The *Project Independence Report*, a cumbersome 781-page document, accompanied by over 20 volumes of background reports, represented the federal government's attempt to respond to these needs. It supplemented a fairly comprehensive quantitative assessment of the domestic policy alternatives and the domestic consequences to be expected from following these policies with a more modest, qualitative assessment of their consequences on the international situation. This chapter deals both with the substantive content of that report and the bureaucratic politics that preceded its publication.

Bureaucratic Politics and the Report

The report is a study in contrasts. It presents a wealth of information and contains a large number of interesting and important insights. At the same time it is filled with internal contradictions, ambiguities, and is extremely difficult to read and comprehend. These characteristics result directly from conscious decision not to compromise the publication date for the sake of issuing a more polished report.

As described above, the modeling effort itself took almost the full time allotted for the production of the report. In fact, there was not even enough time to complete all of the analysis, much less write the report. Therefore, an explicit decision had to be made: either the publication of the report would be delayed, to permit time for completing the analysis and rewriting, or it would go to the publishers in the best form that could be accomplished within the existing deadlines. The costs of delay were political; there was an enormous pressure for an energy plan and the report could assuage that pressure somewhat by providing evidence of progress toward that goal. The main cost of not delaying was the possibility that the shortcomings of the mode of presentation would be transferred, in the minds of public officials, to the underlying analysis itself and would, as a result, undermine the power and effectiveness of this analytical capability as a basis for formulating energy policy. The decision was made that the costs of delay exceeded the costs of rapid publication and the report was published in its unadorned state.

Strict adherence to the deadline led to other costs as well. The success in pulling large numbers of personnel from other government agencies into the effort had been seen as a way not only to enrich the analysis by broadening the base of expertise, but also to increase the receptivity of the bureaucracy toward the modeling system by giving other agencies some stake in it. This would, it was hoped, minimize the problems of coordination and increase the support for using comprehensive models within the executive branch.

This hope, too, was a casualty of the short time allowed for the analysis. The draft report could not be written until the analysis was finalized. Most of the analysis used in the report was not finalized until very shortly before the publication date. As a result, government agencies, other than the Federal Energy Administration (FEA), were given only about a week to review the draft document before it was published. Because of the phenomenon of bureaucratic time compression, discussed in the first chapter, this was an intolerably short deadline om the point of view of the other agencies. Their reaction was emphatically negative.[a]

Aside from the unintended, but nonetheless perceived, insult to the other agencies, the necessity for getting the document out caused a problem of coordinating substantive differences among agencies. Most agencies really did not have time to play a major role. One agency, however, the State Department, was actually able to force substantive changes in the international section of the report.[b] Reflecting the internal dissension within the administration, the draft report had been somewhat critical of the International Energy Program (IEP), a State Department pet project. Specifically, it argued that the IEP would actually be detrimental to the United States when United States import levels were below

---

[a]This hostility is described in Joel Havemann and James G. Phillips, "Energy Report/Independence Blueprint Weighs Various Options," *National Journal Reports*, 2 November 1974, p. 1653.

[b]This can be seen by comparing the statements made in the draft version about the effect of the International Energy Program on import vulnerability as reported in Havemann and Phillips, "Energy Report," p. 1650 with the statements in the final report.

6 million barrels a day. In that case, the report argued, if an embargo should occur, the United States would have to export petroleum to other nations during the embargo.[1] Since it was quite likely that whatever energy program was chosen to complement the IEP, it could well result in the 1980s in import levels under 6 million barrels of oil a day, this difference of opinion was important. The final version of that section, not surprisingly, finds IEP a praiseworthy plan,[2] although careful readers can still find the validity of the original criticism reaffirmed, by piecing together information buried in the revised charts.[c]

In spite of its literary limitations, the report did provide quantitative answers to a large number of questions. What are likely future import levels and how sensitive are they to the world price of crude oil? What kind of flexibility does the United States government have in controlling these import levels? What are the environmental and economic consequences of implementing import reduction programs? What are the distributional implications of these programs? We now turn to the answers to these questions as provided in the Project Independence Report.

Policy Choice and the Petroleum Sector

The petroleum sector was, of course, a key focal point for the policy deliberations because of its importance in most of the energy problems under consideration. There was substantial interest in gaining some knowledge about the degree of flexibility that was available to the United States government, how this flexibility was affected by the Organization of Petroleum Exporting Countries' (OPEC) decisions, and, finally, how the picture changed over time.

The quantitative estimates from the FEA energy model needed to form these judgments are given in Table 5-1. There are several important implications of these numbers. The first, and perhaps most significant, is the overriding importance of the world price for crude oil. Nowhere is this reflected more forcefully than in the import numbers. A return to historical oil prices (roughly $4 a barrel in real terms) could be expected to lead to marked increases in energy consumption in the period to 1985 with petroleum capturing an even larger share of total United States energy consumption than it held prior to the embargo. Since the competition with low-cost imports would simultaneously discourage domestic production, by 1985 imports could be reasonably expected to account for about 72 percent of all petroleum consumption.

This situation of rapidly increasing foreign dependence can be contrasted with the 1985 base case forecast for $11 oil (i.e., continued high prices). The

[c]Compare Table VII-12 on p. 376 with Table VIII-6 on p. 363. This comparison reveals that if an embargo were to occur when the United States is importing 6 million barrels a day or less the United States shortfall would actually increase by participating in the IEP, because it would have to share its domestic supplies with other nations.

Table 5-1
Project Independence Projections of Key Petroleum Variables Under Alternative Government Strategies

Key Variables and Strategies	Actual 1972	$4 World Price		$7 World Price		$11 World Price	
		1977	1985	1977	1985	1977	1985
Total Annual Energy Consumption (Quadrillion of BTUs)							
Base case	72.1	N.A.	118.3	82.6	109.1	78.9	102.9
Conservation	72.1	N.A.	N.A.	80.1	99.2	76.9	94.2
Accelerated supply	72.1	N.A.	N.A.	N.A.	109.6	78.9	104.2
Combined	72.1	N.A.	N.A.	N.A.	99.7	N.A.	96.3
Percent Annual Energy Consumption from Petroleum Products							
Base case	45.7%	N.A.	51.0%	45.8%	43.9%	43.3%	36.9%
Conservation	45.7	N.A.	N.A.	44.6	43.0	42.1	35.6
Accelerated supply	45.7	N.A.	N.A.	N.A.	43.4	43.3	36.5
Combined	45.7	N.A.	N.A.	N.A.	41.5	N.A.	37.0
Domestic Consumption of Petroleum Products (Millions of Barrels per Day)							
Base case	16.5	N.A.	29.8	18.7	23.9	17.0	16.1
Conservation	16.5	N.A.	N.A.	17.6	21.2	16.1	14.0
Accelerated supply	16.5	N.A.	N.A.	N.A.	23.7	17.0	15.5
Combined	16.5	N.A.	N.A.	N.A.	20.6	N.A.	14.4
Imports of crude oil and petroleum products (Millions of Barrels per day)							
Base Case	4.7	N.A.	21.4	9.2	12.4	6.6	3.3

Conservation	4.7	N.A.	N.A.	8.2	9.8	5.7	1.2
Accelerated Supply	4.7	N.A.	N.A.	N.A.	8.5	6.6	0.0
Combined	4.7	N.A.	N.A.	N.A.	5.6	N.A.	0.0
Domestic production of crude oil and petroleum products (Millions of Barrels per Day)							
Base case	11.8	N.A.	8.4	9.5	11.5	10.4	12.8
Conservation	11.8	N.A.	N.A.	9.4	11.4	10.4	12.8
Accelerated supply	11.8	N.A.	N.A.	N.A.	15.2	10.4	15.5
Combined	11.8	N.A.	N.A.	N.A.	15.0	N.A.	14.4

Source: These numbers were taken from the FEA model outputs. Most of them have been reprinted in the *Project Independence Report*, appendix pages 33-49 for the first three variables, and p. 318 for import levels. The domestic production figures are computed by subtracting imports from domestic consumption.

higher prices of competitive petroleum imports were estimated to diminish demand and encourage domestic production to such an extent that the situation, in terms of import dependency, was reversed from the $4 case. In the $11 case only about 20 percent of petroleum consumption was estimated to come from imports by 1985. The clear implication of these numbers is that the problem of import vulnerability and the necessity for action are quite dependent on actions taken by the OPEC nations.

These estimates also point out that the short-run import reductions, in response to increased world prices, are much less than the long-run impacts. This reflects the well known fact that short-run domestic demand and supply curves are much less price responsive than their long-run counterparts.

The various policy strategies, as articulated in Project Independence, exercise a less dramatic impact on the import dependency projections. For example, unlike the changes in world crude prices, none of the strategies is particularly successful in causing the share of petroleum in total energy consumption to fall. The accelerated supply strategy, for example, which contained, among other programs, an accelerated program of nuclear development, resulted in only small reductions in the consumption levels of petroleum by 1985. The rather small size of this reduction occurred because the nuclear power plants could be expected to replace coal-fired power plants, rather than oil-fired plants. As a result it did not have much affect on petroleum consumption. Import levels, however, were expected to be affected by an accelerated supply strategy because of the increased production from the Naval Petroleum Reserves and the Outer Continental Shelf. In fact, by 1985, if the price of world crude were to remain at $11 a barrel in real terms, a strategy of accelerated supply could be expected to yield self-sufficiency in petroleum.

Unlike accelerated supply, a conservation strategy would be expected to reduce total energy consumption, relative to a base case, as well as petroleum consumption. This reflects the orientation of the strategy toward a reduction in energy use rather than toward encouraging substitutions of one energy source for another. Almost none of this reduction would come out of domestic production because, at the prices considered, most domestic oil can be produced for less than the import price. As a result petroleum consumption reductions are matched by petroleum import reductions.

The final question addressed by Project Independence was the role of unconventional sources in providing substitutes for natural crude oil in the future. To the extent that good substitutes are likely to be available in large quantities at low prices the case for limiting domestic production is weakened. Conversely, the absence of such substitutes would strengthen the case.

The development of supply curves for unconventional sources is a risky undertaking because the technology and, hence, the production costs are in a state of rapid change. Increased interest in an unconventional source leads to increased research and development, which can, in turn, lead to better technical and economic efficiency in the production and use processes. The analyst has to guess where all these trends will lead.

The production of synthetic liquid fuels from coal was estimated to be very small by 1985. This rather small contribution was mainly due to development and production time lags. Synthetic liquid fuels were expected to be economically competitive with $7 oil on the Pacific Coast, but could only compete with $11 crude oil elsewhere.[3] This geographic difference in economic viability resulted from differential access to the relatively inexpensive Western coal.

Shale oil, a product extracted from oil-bearing marlstones, primarily in Colorado, Wyoming, and Utah, represents the largest concentration of organic deposits known to exist in the world.[4] Over 600 billion barrels of high-grade reserves are known to exist. Unfortunately, there are serious constraints on the development of this vast resource. Shale oil was estimated to be competitive to $7 crude oil, but, at this price, only a quarter of a million barrels a day could be produced by 1985.[5] Production levels could be much higher (1.6 million barrels a day by 1990) with an accelerated development program,[6] but the undertaking of such an accelerated program of development could be expected to cause significant pollution problems and to exacerbate local water scarcity problems.[7]

Another substitute, of particular interest because of its abundant supply and generally desirable environmental characteristics, is solar energy. The Solar Task Force estimated that by 1985 the various forms of solar energy could supply between 1 and 2 quadrillion BTUs of energy if the world price of oil were to remain at $11 a barrel.[8] This represents only about 1 percent of the projected total domestic energy consumption. However, it was also estimated that the contribution that could be made by solar energy could reach 39 quadrillion BTUs by the year 2000,[9] which would represent a significant portion of total demand. If these estimates are accurate, the long-term potential for solar energy is large if world prices remain at or above current levels.

The Estimated Impacts of Energy Policy on Subsidiary Objectives

The main subsidiary objectives to be achieved simultaneously with the energy objectives, to the extent possible, involved the maintenance of a strong, stable economy, an equitable distribution of the costs and benefits of energy policy and the protection of the environment. The quality and quantity of information bearing on these objectives was limited by time, the state of the art, and the unavailability of key data. Nonetheless some limited information was presented in the report.

The first area of interest was the relationship between energy policy and economic growth. There were two main concerns suggesting that domestic attempts to reduce imports would result in lower economic growth. The first of these held that the substitution of high-cost domestic energy for low-cost foreign energy would raise domestic prices, lower real incomes, and put the United States in a noncompetitive world trade position. The second concern was that the required investment for energy could not adequately be met in existing capital markets without slowing the growth of the economy.

The report concluded that within the range of likely world prices ($7 or $11) neither of these concerns appeared valid. The first point was refuted by pointing out that since most domestic reserves of oil could be produced at prices under $7 a barrel, increasing domestic production simply led to the substitution of cheap domestic oil for expensive foreign oil.[10] Because the domestic price would be largely determined by world price, this would not necessarily result in lower prices, except when the domestic production was more than adequate to satisfy all demand at lower prices. The fear that substituting domestic for foreign oil would slow economic growth, while valid for very low, world crude oil prices, was no longer true at the prevailing world prices. If the cartel were to completely collapse, however, and the world price of crude would fall to $4 a barrel, then exploiting domestic resources would raise prices because production would have to be subsidized by some sort of price guarantee.

While higher world prices tend to diminish the importance of this first concern, they increase the importance of the second concern. With higher energy prices the demand for energy investment would be much higher and its impact on capital markets would potentially be much larger. To quantify this impact the relationship between historical and future investments for both the business as usual and accelerated development strategies were examined. The accelerated development strategy was found to lead to marginal higher investments, both in the energy sector and in total,[11] and these were expected to cause somewhat higher long-run interest rates. The report also found, however, that these levels would not deviate from historical norms by enough to affect economic growth adversely.[12] On the basis of this evidence the report concludes that there is no long-run conflict between the pursuit of energy objectives and the desire to maintain economic growth.

The distributional effects of energy policies were considered both in terms of spatial differences and in terms of differences among socioeconomic groups. The spatial distributional effects were considered both in terms of the effects on regional energy costs and on regional earnings.

Regions have different energy mixes and face different energy costs because of the nonuniform distribution of energy resource endowments across the country. This spatial inequality in endowments leads to a spatial inequality in delivered costs. Since the energy strategies would lead to the development of different resource endowments at different times, this raised the question of the extent to which the various strategies would change the geographic distribution of relative energy costs. Increases in energy cost disparity among regions could have important long-term implications for industrial location and, hence, regional growth. The report found that while all policy strategies did tend to increase the disparity among the regions, compared to the level that would have existed in the absence of the strategies, this increase was likely to be rather small.[13] This was due to the fact that while all three strategies were estimated to reduce BTU prices in all regions, relative to what they would have been in the absence of policy, New England, the region with the highest energy costs, received the smallest reduction.[14] New England was also identified as being the region most affected by changes in the import prices of petroleum, since it was the region most dependent on imports.

The distribution of regional earnings was not projected to be affected very much by the various possible energy strategies. The particular boom-town developments, which could be expected, were too small in terms of total state earnings to make much of a difference on the state as a whole. The major exceptions were Alaska, which could be expected to grow significantly faster than the national rate, particularly if an accelerated supply strategy were followed; Indiana and Michigan, which could expect retarded growth from slower automobile sales under a conservation strategy; and West Virginia, which would experience reduced earnings from an accelerated supply strategy because coal production would be lowered by an increased emphasis on nuclear power.[15]

The socioeconomic impacts were estimated by examining the effects of the various strategies on the real income of various socioeconomic groups. This required an examination of the effects on the distribution of income and of the incidence of price effects on the distribution of expenditures. The data were generally not available to provide an explicit quantification of the effects on real incomes, but an examination of the available data suggested some qualitative conclusions. Projections of the functional distribution of income revealed that labor's share had fallen somewhat during the embargo and subsequent period of unemployment.[16] Labor's share was expected to recover by 1985 to its 1970 level as the economy regained strength and real wages increased. However, since energy investment was expected to drive interest rates up, the share of interest payments in national income could also be expected to increase. The former effect would benefit lower income people and the latter primarily, but not exclusively, the well-to-do.

Higher energy prices were found to be quite regressive. The direct expenditures on energy as a percentage of income fell with rising income.[17] The indirect effects of higher prices (i.e., the effects of energy price rises on the prices charged for other goods) are more proportional to income.[d] Putting these esti-

---

[d]This conclusion was independently reached in Bradley W. Perry, "The Short Run Consequences of Increased Energy Cost," *Energy Systems and Policy* I (1974): 70; and in Ford Foundation Energy Policy Project, *A Time to Choose: America's Energy Future* (Cambridge: Ballinger, 1974), p. 126.

mates together yields the conclusion that energy price increases are, on balance, somewhat regressive. To the extent that demand elasticities are lower for lower income groups, as has been suggested,[18] the regressivity of energy price increases would be even higher than estimated.

The final subsidiary objective involved the environment. As described in Chapter 6, the environmental analysis had to be accomplished on a very aggregate level. This made the analysis decidedly less useful because the main environmental problems are local, not regional, in nature. For example, the air pollution problems associated with the production of oil shale would not appear very great for Colorado as a whole, but for the local area in which the shale was being processed they may be quite significant. The conclusions in the report, therefore, are based on a consideration of aggregate regional emissions of pollutants and regional land use requirements. To provide some basis for comparison, the 1972 actual figures were provided along with estimates for each scenario.

On a national basis it was found that, with the exception of organic pollutants (oil spills) and thermal discharges, water pollution from energy development could be expected to fall from 1972 levels under all strategies.[19] This, perhaps, unexpected, finding is caused by the anticipated increase in the use of control technologies in response to rather stringent existing federal law. This increased adoption of control equipment would offset the increase in uncontrolled emissions caused by increased energy development. Oil spills could be expected to increase over 1972 levels in the accelerated development strategy in Alaska and the Northeast due to Outer Continental Shelf development.[20] Air pollution problems could be expected to increase over 1972 levels for particulates, nitrogen oxides, and aldehydes, while generally decreasing for hydrocarbons, sulphur oxides, and carbon monoxide.[21] The decreases in these last three air pollutants also result from the increasing impact of federally mandated pollution control standards. The sulphur oxide emission levels could be expected to increase, however, if a large switch from oil burning to coal burning power plants were initiated.[22]

Solid waste pollution could be expected to increase markedly from an accelerated development strategy due to increased oil shale development, mainly in the mountain states.[23] In perhaps its most striking environmental finding the report concludes that the amount of land allocated to energy production, conversion, and use was expected to rise markedly over its 1972 level. This increase would be felt in all regions and was relatively insensitive to changes in the world price of crude and to the choice of strategies.[24] Land use planning and energy planning were inevitably linked.

The final environmental finding concerned the relationship between pollution and the world price for crude oil. Compared to a referent case of continued high world prices a fall in world oil prices could be expected to lead to an increase in consumption, a fall in domestic production, and an increase in imports. The higher consumption and imports would cause more pollution problems while the lower domestic production would cause less. The report concludes that the latter effect would outweigh the former and therefore domestic pollution problems would be less intense with a lower world price for crude.[25]

Identification of Constraints

The constraint analysis, one of the weakest sections of the report, was conducted by comparing projected requirements for labor, water, capital, materials, and equipment as implied by the FEA energy model with projected supplies. The future available supplies were naively estimated by projecting GNP and making some assumptions about the share of GNP allocated to these supplies.[26] Adequacy was then determined by comparing projected demands with projected supplies.

Several potential constraints were identified by the analysis. The capacity to implement the accelerated supply strategy was estimated to be limited both in the short run and in the long run by the availability of mobile drilling platforms. There was also estimated to be a shortage of drilling rigs (used for exploratory drilling on and off shore) in the short run.[27]

Water was also judged to be a potentially serious constraint. A particularly serious conflict over water in the upper Colorado region and in Missouri was anticipated from shale oil development and the production of synthetic fuel from coal.[28] The conflict arises because these processes use large amounts of water, but the regions in which the deposits of coal and oil shale are found are relatively arid regions.

In addition to the problems associated with financing the expected aggregate level of investment, as discussed above, the financial task force also looked at each of the energy sectors to establish the possibility of financing the required capital expansion. Although the examination of each energy sector revealed that financing requirements could apparently be met easily by the coal industry, and the oil and gas industry,[29] it did find one sector that could potentially have financial problems—electric utilities. This difficulty resulted both from the fact that electric utilities are regulated, so internal funds would not normally rise fast enough to finance the expansion,[30] and, further, from the fact that the switch to nuclear baseload plants required larger capital outlays.[31] These nuclear plants were forecast to reduce operating cost because the fuel was expected to be cheaper, but they would require larger capital expenditures to construct. The report also acknowledges, however, that a much less gloomy report would occur if electricity turned out to be more sensitive to price than estimated in the Project Independence analysis.[32]

Energy Objectives and Strategy Choice

Having examined some of the more important individual pieces of the analysis, it is time to weave them together with additional information into an evaluation of the impacts of alternative strategies on the three main objectives of energy policy.

The first major problem area to be addressed is import vulnerability. What was the nature of import vulnerability in the long run and short run? How much control could be exercised over import vulnerability? How serious a problem was import vulnerability?

The key determinants in the import vulnerability problem are the level of imports, the proportion of total petroleum consumption formed by these imports, and the proportion of total energy contributed by petroleum. The key estimates were presented earlier in Table 5-1. An estimate of the nature of the petroleum dependence in the absence of policy, except for the deregulation of natural gas prices and the decontrol of oil prices, can be seen by an examination of the base case in Table 5-1. The importance of the world price of crude oil dominates the picture; it affects the importance of petroleum in total energy consumed, the level of petroleum demand and the level of domestic production. Since higher prices tend to move all of these variables in directions that lower imports, the dramatic effect of these very large price increases should not be surprising. Note that the long-run and short-run responses to higher prices are quite different. This follows from the realization that both the demand elasticities and the supply elasticities are low in the short run, but quite a bit larger in the long run. The net result of this on import dependency is summarized in Table 5-2. The picture that emerges from this table is that rising dependence is inevitable in the period until 1977. This generally is due to rising

Table 5-2
Projected Shares of Total Domestic Energy Coming from Petroleum Imports Under Alternative Government Strategies

Strategy	1972 Actual	World Price of Crude					
		1977			1985		
		$4	$7	$11	$4	$7	$11
Base case	13%	N.A.	23%	17%	37%	23%	8%
Conservation	13	N.A.	21	15	N.A.	20	3
Accelerated supply	13	N.A.	N.A.	17	N.A.	16	0
Combined	13	N.A.	N.A.	N.A.	N.A.	11	0

Source: Computed from Table 5-1.
Note: N.A. = Not available.

demand and falling domestic production. The relative importance of the various Project Independence strategies in reducing import dependence during this period is also clearly evident.

In the long run, however, the picture is quite different. A drop in the world price to preembargo levels would, in the absence of a countervailing United States policy, cause a marked increase in dependency. On the other hand, continued high oil prices could be expected to reduce the dependency level to below its 1972 level. And by 1985 the various strategies gave the government a moderate degree of flexibility in choosing an appropriate degree of dependency; it was not entirely at the mercy of OPEC.

Table 5-2, as useful as it is, does not present the complete picture of import vulnerability. Two issues remain to be covered: (1) what contingency programs could be adopted to complement the basic strategies; and (2) how serious is the import vulnerability problem? With respect to the first point, the Project Independence analysis concluded that a strategic reserve, or stockpile, of petroleum was a very cost effective way to reduce vulnerability.[33] With a certain lack of political foresight, the international assessment also states, "tariffs do not appear to provide an economically efficient means of enhancing supply security."[34]

The report also made a limited attempt to assess the likelihood of alternative OPEC pricing strategies. Drawing upon work done by the Organization for Economic Cooperation and Development[35] the international assessment task force estimated the 1985 production capacity for the OPEC nations could be about 53.1 million barrels a day.[36] This contrasts to a projected world import demand of about 29.4 million barrels a day with continued high prices.[37] This would lead to a projected excess capacity of some 46 percent for the producing group as a whole.

The report is silent on which pricing strategy would maximize OPEC revenues, but it provides enough information to conclude that through 1985 revenue is maximized for the period as a whole by maintaining high prices.[e] The report also points out, however, because of differences in revenue absorption these potential excess capacities are not likely to be equally shared by all OPEC members. If all those members of the cartel who can absorb the additional revenues are allowed to maximize their production, with the remaining members sharing in the excess capacity, these latter nations would have a 52 percent excess capacity. If OPEC were to settle for a $6 price (corresponding roughly to a $7 price delivered in New York), these nations would only have a 12 percent surplus.[38]

The report therefore concludes that, although the present value of revenue

[e]Federal Energy Administration, *Project Independence Report* (Washington: U.S. Government Printing Office, 1974), p. 358. The figures in Table VII-3 of the report can be used to see that by 1985 the revenue is only marginally higher from the lower price. Since higher prices in the earlier years lead to substantially higher revenues, it is clear that the discounted revenue stream for the higher price must exceed that for the lower price.

for the cartel as a whole would be maximized by continued high prices, the stability of the cartel would be subjected to considerable pressure because of difficulties in allocating the potential production cutbacks among the members.[39] This general view seems to have been shared by other observers, who have subjected the question to a more careful scrutiny.[40]

With respect to the balance of payments problem the Project Independence analysis restricted itself to that part of the problem for which it could provide good quantitative estimates, the expected dollar outflow for oil. These calculations reveal that the dollar outflows for oil would decline over time for continued $11 oil or would rise slightly with a $7 oil.[41] In either case expressed as a percentage of real gross national product these outflows would decline. From the United States point of view, any of these dollar outflow levels could be managed. Financing the deficit for the United States was not expected to be a problem in the short run because a fairly large proportion of OPEC investment was expected to flow back into United States markets. In the longer run, rises in exports to the OPEC nations were expected to bring the trade account back into balance. The main problem was expected to occur in the short run for nations unable to finance their deficits because they would not attract enough foreign investment. This, it was expected, could be handled by focusing on the institutional changes needed to facilitate recycling of the petrodollars.[f]

The *Project Independence Report*, in deference to the State Department's jurisdiction in the area, had very little to say about foreign policy initiatives in general as a solution to the energy problem. Its only comment on the subject was to point out that a strategy of import source diversification was not a viable alternative within the context of the International Energy Program because, during an embargo, supplies would be shared among the consuming countries anyway.[42] The United States was explicitly choosing a policy of cooperation with consuming nations, not competition for scarce, secure sources.

Notes

1. Joel Havemann and James G. Phillips, "Energy Report/Independence Blueprint Weighs Various Options," *National Journal Reports*, 2 November 1974, p. 1653.

2. Federal Energy Administration, *Project Independence Report* (Washington: U.S. Government Printing Office, 1974), pp. 369-76.

[f]This characterization of the problem was fairly widely held. See, for example, Edward Fried, "Financial Implications," in Joseph A. Yager and Eleanor B. Steinberg, *Energy and U.S. Foreign Policy* (Cambridge: Ballinger, 1974), pp. 277-310; and U.S., Congress, House, Ad Hoc Committee on the Domestic and International Monetary Effect of Energy and Other Natural Resource Pricing of the House Committee on Banking and Currency, *Balance of Payments to Higher Oil Prices*: Managing the Petrodollar Problem; 93rd Cong. 2d sess., 1974.

3. Federal Energy Administration, *The Final Report of the Synthetic Fuels Task Force* (Washington: U.S. Government Printing Office, 1974), p. 104.

4. Federal Energy Administration, *Potential Future Role of Oil Shale: Prospects and Constraints* (Washington: U.S. Government Printing Office, 1974), p. 89.

5. Ibid., p. 63.

6. Ibid.

7. Federal Energy Administration, *Project Independence Report*, p. 134.

8. Federal Energy Administration, *Solar Energy: Final Task Force Report* (Washington: U.S. Government Printing Office, 1974), p. I-1.

9. Ibid.

10. Federal Energy Administration, *Project Independence Report*, p. 317.

11. Ibid., p. 320.

12. Ibid.; and Federal Energy Administration, *Financing Project Independence, Financing Requirements of the Energy Industries, and Capital Needs and Policy Choices in the Energy Industries* (Washington: U.S. Government Printing Office, 1974), p. 23.

13. Federal Energy Administration, *Project Independence Report*, p. 331.

14. Ibid., p. 329.

15. Ibid., p. 333.

16. Ibid., p. 335.

17. Ibid., p. 341.

18. Dorothy K. Newman and Dawn Day, *The American Energy Consumer* (Cambridge: Ballinger, 1975), p. 90.

19. Federal Energy Administration, *Project Independence Report*, p. 216.

20. Ibid., p. 225.

21. Ibid., pp. 218-19.

22. Ibid., p. 219.

23. Ibid., p. 227.

24. Ibid., pp. 220-27.

25. Ibid., p. 223.

26. Ibid., p. 229.

27. Ibid., p. 247.

28. Ibid., p. 310.

29. Ibid., pp. 289-90.

30. Ibid., p. 287.

31. Ibid., p. 285.

32. Ibid., p. 288.

33. Ibid., p. 403.

34. Ibid., p. 397.

35. Organization for Economic Co-operation and Development, *Energy Prospects to 1985: An Assessment of Long Term Energy Developments and Related Policies: A Report by the Secretary General*, vols. 1 and 2 (Paris: Organization for Economic Co-operation and Development, 1974).

36. Federal Energy Administration, *Project Independence Report*, p. 356.

37. Ibid., p. 357.

38. Ibid., p. 359.

39. Ibid., p. 360.

40. Joseph A. Yager and Eleanor B. Steinberg, *Energy and U.S. Foreign Policy* (Washington: Brookings Institution, 1974), p. 273.

41. Federal Energy Administration, *Project Independence Report*, p. 325.

42. Ibid., p. 402.

6

The Design, Control, and Use of the Project Independence Modeling System

As important as the *Project Independence Report* was in assembling a large amount of information, by itself it could not fulfill the information needs of the policy process. While it characterized a large number of policy possibilities, it was inevitable that it would fail to address the particular concerns of various members of the policy process. For this reason an attempt was made to make the modeling system used to generate the *Project Independence Report* available to the policy participants so that they could assess the consequences of scenarios of particular interest to them.

This chapter deals not only with the design and control of this modeling system, but also with the less tangible, although no less important, subject of the environment in which the system was developed and used. It provides the answers to several questions. What problems were encountered and successes achieved in developing this modeling system? How was such a large-scale effort controlled? What information channels were used to feed the analysis into the policy process? How effective were they? What affect, if any, did the system have on the policy outcomes at various stages of the process?

Management of the System

The focal point for the analysis was the Federal Energy Administration, although substantial contributions were made by a large number of other public organizations and private consulting firms. During the time that the modeling system was being developed and used the top management position at the FEA was occupied by three different people.[a] Since this rotation, and the sequence in which it occurred, affected both the development and the use of the modeling system, the chapter begins with a description of these men and the role they played.

Following the onset of the Arab oil embargo the need to concentrate and coordinate the government effort under strong leadership led to the establishment of the Federal Energy Office (FEO) under the President. It was headed by William Simon, who also retained his former job as deputy Treasury secretary

---

[a]One other person was nominated to be an administrator, Andrew E. Gibson, but because of the uproar that greeted his nomination, owing to his close connections to a large oil firm, he withdrew his nomination on November 12, 1974, less than two weeks after it had been submitted.

while serving as the FEO administrator. His deputy at FEO was John Sawhill, who moved to that position from the Office of Management and Budget. This transition from the previous system of powerless committees to deal with energy was important for several reasons. Simon had a reputation, an accurate one, for getting things done. One of his first accomplishments was to mobilize a staff, obtained largely from other agencies. For example, he wasted no time in wrestling the Office of Oil and Gas Policy and some key staff, including Eric Zausner, the eventual Project Independence analysis manager, away from the Department of Interior. The new agency also drew some key staff from the Office of Management and Budget, the Departments of Labor and Commerce and the Environmental Protection Agency.

The new organization and its staff were, in the early months, necessarily preoccupied with managing the petroleum allocation program initiated in response to the Arab oil embargo. Little long-range planning was accomplished. But then the embargo was lifted and the departure of George Shultz as secretary of the Treasury opened up that position for Simon, which he took. Sawhill was then elevated to the vacant administrator's job.

Sawhill's ascension also represented an important change. As a former academic and a relative neophyte in the political world, he was committed to a rational decision-making process. He wanted all alternatives considered and weighed. He wanted the analytical effort to play an important role and, hence, he gave it a great deal of financial support, if not much of his attention.

The work began in earnest, under Eric Zausner, the assistant administrator for policy and analysis, in April 1974. Zausner was a key figure both in the development of the modeling system and its use, because he served as an effective link between the decision makers and the analysts. He made sure the system would be controllable and timely.

A key procedure in achieving model controllability was the meeting held at 3:00 P.M. every day during the model development phase. This meeting was attended, among others, by Zausner; Bart Holaday, the deputy assistant administrator for analysis; Bill Hogan, the chief architect of the supply and integration models; Bruce Pasternack, the deputy program manager and the head of the Policy Evaluation Task Force; David Wood, the head of the demand modeling group; Bill Stitt, the consultant in charge of deriving the oil and gas supply curves; and Al Cook, the head of the economic impact assessment. Through constant exposure to this very complex model and its behavior under a variety of circumstances this group was able to gain a thorough understanding of its inner workings. This was invaluable later in explaining particular results to the principals in the policy process and being able to counter incorrect charges that this detail or that detail was not adequately considered by the model. Zausner and the other policy participants learned to use the power of the model while being careful not to oversell its capabilities.

Realizing that only timely analysis would ever be used, Zausner sought to insure that the *Project Independence Report* was available approximately when expected and that the models developed would be capable of providing rapid responses to questions about alternative policies. The objectives were achieved by making it clear that deadlines were rather sacred. Rigid adherence to deadlines was accomplished through the liberal use of outside consultants to supplement the staff, through a de facto extension in the length of the working day to something frequently approximating 14 or 15 hours a day for the principals and by sacrificing some of the model design characteristics when it became clear they would take too much time.

While the Sawhill-Zausner team was effective in developing the system, they were not effective in integrating it into the decision-making process. Sawhill was not an administration insider. He did not possess the power or authority to provide an effective channel to feed the analysis into the policy process. The climate soon changed, however, because Sawhill resigned under pressure on October 29, 1974. Somewhat prior to this the Energy Resources Council, headed by Secretary of the Interior Rogers C.B. Morton, was formed to coordinate and consolidate the energy policy formation process. The executive director of the group was Frank Zarb, from the Office of Management and Budget. Zarb had held an important position at FEO during the Arab oil embargo and had developed a very good working relationship with Eric Zausner during that time. This relationship was organizationally reinforced when Zarb, who retained his role in the Energy Resources Council, became the third Federal Energy Administration (FEA) administrator and chose Zausner to be one of his deputies.

After Zarb entered the picture, the modeling system was used in the policy making process for the first time. Zarb was an administration insider. He was not only a key figure in the Energy Resources Council, who had direct access to and the confidence of the President,[1] he also was highly regarded on Capitol Hill.[2] He appreciated the value of analysis for developing politically saleable policy packages and for selling them. With the formation of the Zarb-Zausner team, the information channel in the executive branch between the modeling system and the policy process was complete. The channels to the legislative committees considering the key pieces of energy legislation were soon to follow, albeit less formally and less effectively.

The System Development Process

In April the process of operationalizing the conceptual framework began. Several questions had to be answered: What was to be the planning horizon? What level of spatial aggregation would be attempted? Given the time constraints on the analysis what was the best type of system to be used? What information sources would be needed and where would they come from? How would uncertainty be treated? What kind of validation procedure could the effort be subject to? How could this mammoth exercise be kept under control? We consider here the answers provided to each of these questions.

The choice of the planning horizon was an important one. The nature of the energy situation could change quite drastically over the next 25 years with the development of new fuels, the discovery of more old ones, and changes in how energy is used. As interesting and important as these longer run problems are, projections of that length would necessarily be highly conjectural. Initially it was decided that the models would cover the period up to 1990. This period was long enough to capture the initiation of the longer run changes in the system that could be expected in response to policy, including the arrival on the scene of the newer sources of energy such as solar and synthetic fuels, but short enough that most of the major forces could be currently identified. As one of the concessions to deadlines, the model horizon was subsequently shortened to 1985 and the responsibility for analyzing the post-1985 period was given to an Energy Research and Development Task Force.

The model could either be a national model, concentrating strictly on national aggregates, or it could be more ambitious and deal with a finer spatial representation. The benefits of going to a regional level of detail were many. This would permit the characterization of the transportation system and its role in energy supply and demand. Transportation was expected to be important both as a potential constraint on shipments (e.g., the Alaskan pipeline) and as a significant component of the cost of energy (particularly for shipments from Alaska). Using the region as the basic spatial aggregate also allowed the provision of information about the requirements for new facilities of particular types (e.g., pipelines, refineries, power plants, etc.) in each of the regions. This would provide an important point of departure for further investigations of the specific localities where the facilities might be located. The costs of gaining this information were the computational complexity it would add to the model and the unavailability of complete regional data. The decision was made to use the region as the basic spatial unit for modeling and this decision survived the deadline crunch.

The most important decision to be made was the choice of the modeling techniques to be employed. There were several criteria the system had to satisfy. It should reflect the price sensitivity of the energy system. It should be a general equilibrium system, insofar as possible. It had to be run at FEA by FEA personnel in order to insure controllability and timeliness. It had to be operational in time to use for writing the *Project Independence Report*. Given the tight time constraints it was clear that what was needed was a modularized system, resting heavily on existing models, in which each of the components could be developed simultaneously in coordination with the other components.

The state of the art provided a few alternatives,[b] but the best approach seemed to be a fusion of an econometric demand system with a large-scale linear programming model of the supply and transportation networks. Linear programming had already been used rather extensively to model various components of the energy sector, and, therefore, the state of the art was fairly well advanced.[3]

In terms of how it worked it is easiest to think of the system as performing five major tasks: (1) deriving energy demand, by fuel, and the relationship of demand to policy, (2) deriving energy supply, by fuel, and its relationship to policy, (3) developing a method to derive the equilibrium prices that balance supply and demand, (4) developing a series of impact models to trace out the implications of particular energy configurations on the environment, on the economy, and on the international petroleum market, and (5) identifying and incorporating any constraints that might prevent an unconstrained equilibrium from actually occurring. The nature of this model may well have contributed to the positive reception it later received. The system was not designed to assign a unidimensional ranking to all policy packages as would be done, for example, by a cost-benefit analysis. Rather it traced out the likely implications of policy packages in terms of many national indicators; the choices were left up to the decision makers. Because of this, the system represented less of a threat to the politicians who used it, although in the process it conveyed less information to them.

The information requirements of the modeling system were immense. Separate task forces were set up to obtain the data for each of the major fuel supply curves and constraints. These task forces were generally made up of a variety of government experts drawn from many different agencies with substantial staff support from outside consultants. The chairpersons of these task forces were generally from the agency with the main jurisdiction over the particular fuel or constraint being examined.

There were many sources of uncertainty in attempting to model likely outcomes. These were handled in different ways for different sources of uncertainty. One way was to run the entire system for a range of the possible uncertain events. This, for example, was the approach taken in handling the uncertainty associated with the world price for crude oil. The problem with this approach, when used for all forms of uncertainty, is that it is expensive and it yields more information than is easily comprehended. Most other uncertainties

---

[b]The state of the art as contained in the literature as of 1973 is described in D.R. Limaye, R. Ciliano, and J.R. Sharko, "Quantitative Energy Studies and Models: A Review of the State of the Art" (Final report submitted to the Council on Environmental Quality No. EQC-303, January 1973). A somewhat more accessible description can be found in Philip K. Verlager, Jr., "An Econometric Analysis of the Relationships Between Macro Economic Activity and Consumption," in Milton F. Searl, ed., *Energy Modeling: Art, Science, Practice* (Washington: Resources for the Future, 1973), pp. 64-102.

were handled using "best guess" point estimates. For example, the prices at which new technologies would become economically viable, and the production possibilities that could be expected from these technologies at those prices were supplied by the task forces as point estimates. Although some sensitivity analysis was performed, and recorded in the task force reports, this information was not used in the modeling system.

The validation of comprehensive planning models is as difficult as it is important. The ultimate judgment on their validity cannot be made until the time covered by the planning horizon has elapsed. However, if policy is to be based on these models they must be subjected to some validation procedure.

The validation procedure initiated by FEA consisted of two parts. The first part consisted of pouring over all the input data and checking it for compatibility with the rest of the data (e.g., insuring the use of consistent units) and for mistakes (e.g., recalculating any suspicious numbers). This tedious, but vitally important, effort was handled by a small staff of FEA personnel, assisted by a battery of consultants. It afforded protection from that old computer adage "garbage in, garbage out."

The second validation effort was directed toward the methodology and the results generated by that methodology. The FEA assembled advisory committees consisting of representatives from universities, industry, and government. These committees were given a mandate to assess the methodology and the results. The results of their deliberations were submitted in a report, which was simultaneously made available to the top management of FEA and the analysts. Generally supportive, these reports were useful in refining the system. Their main role, however, in retrospect, was to give a green light to the work that was going on.

A separate validation effort was started by the Office of Management and Budget (OMB). OMB, somewhat concerned over the possibility of being overwhelmed by this technical apparatus being assembled, commissioned the National Science Foundation to conduct an independent review of the *Project Independence Report* and the modeling system behind it. These reports,[4] also quite supportive of the methodology, occurred too late in the process to be of much help in model development, but they were probably significant in gaining support for the effort within OMB.

The implementation of this system was handled through an elaborate structure of task forces involving participants from 23 government agencies and consulting firms too numerous to count.[c] Most of these task forces handled either the development of one of the model modules or provided input data to the system. The Policy Evaluation Task Force, which reported directly to Zausner, coordinated the effort and handled the issue paper process.

The issue paper process was designed to anticipate key policy issues that

[c]The names of these task forces and their members can be found in Federal Energy Administration, *Project Independence Report* (Washington: U.S. Government Printing Office, 1974), pp. 324-33.

emerged as the analysis was going on. These issue papers could be originated either by the analysts or by the top management of FEA. These papers served to link policy concerns to the modeling effort. As such they helped in defining the scenarios that were evaluated in the *Project Independence Report*, and in writing the report itself.

One such scenario, the business as usual strategy, was supposed to represent what would happen if no new policy initiatives were taken other than removing the existing price controls for domestic crude oil and implementing the deregulation of natural gas. The inclusion of these last two elements in the strategy was a controversial decision. The decision was made on both political and modeling grounds. The modeling grounds were that, since the energy model was basically a competitive equilibrium model, it would be difficult to include these controls without a significant amount of exogenous calculation. In essence, the model would have to be tricked into achieving a solution it would not normally produce. Given the fact that any scenario embodying controls would be subject to a considerable amount of ad hoc adjustments and would, therefore, be suspected of being dominated by political considerations, the decision was made not to generate any solutions containing continued controls on crude oil and natural gas prices. This eliminated one difficult modeling task that allowed the analysts to concentrate on the other strategies, but it certainly left a gap in the usefulness of the output for two major policy issues.[d]

Structure of the System

The modeling system that was developed was not the fully simultaneous system that would have been desirable. The feedback linkages between the energy model and the economic model were never completed. In addition, the constraints, other than the transportation constraints, were not able to be included as part of the modeling effort in time to meet the report deadline.

The demand system was composed of four submodels: (1) a price submodel to provide an initial time trajectory for important price variables, (2) a macroeconometric model to provide time trajectories for the major independent variables such as income, housing starts, etc., (3) an econometric simulation model to provide national energy quantity forecasts contingent upon the information provided by the macroeconometric and price models, and (4) a share model to allocate these quantity forecasts among the census regions. The purpose of this system was to provide a series of quantity forecasts by energy

[d]Some compensation for this was provided by including in the *Project Independence Report* the best guess concerning the effect of continued regulation. See, for example, Federal Energy Administration, *Project Independence Report*, p. 95. This problem was circumvented in the analysis conducted during the policy formation stage by forecasting the short-run effects of controls with a different model. This model could more easily treat the effects of controls.

type for each census region that was compatible with the exogenous price trajectory and to provide a matrix of own price and cross price demand elasticities.

The price submodel took price trajectories for primary energy products (e.g., crude oil) and translated these into price trajectories for energy products (e.g., electricity). Historical retail and wholesale markups were used to form the basis for this relationship.

The macroeconometric model was used simply to provide projections of key exogenous variables through 1985. The use of this model was yet another area where the limited time available compromised the quality of the analysis. The compromise took two forms. In the first place the macroeconometric model used to initialize these demand forecasts was not the same macro model used to characterize the economic impacts of each of the strategies. Second, the same set of exogenous variables was used to generate all the demand forecasts, which would be accurate only if these variables truly were unaffected by the availability and cost of energy. The analytical importance of these compromises was that the income side of energy consumption, which changed from scenario to scenario, was inadequately reflected in these demand forecasts.

These deficiencies were recognized from the outset and procedures were established to circumvent them. The plan was to run through the complete scenario, including getting ex post macroeconometric model forecasts, and then using these to initialize the entire process over again, at least for strategies that exhibited growth paths which deviated significantly from the initializing solution. This plan was never implemented because the first stage of this process was never completed until shortly before the final document went to the printers.

The national demand model represented the core of the demand system. It represented a system of price sensitive, econometrically estimated relationships. The general procedure was to estimate total energy demand for each consuming sector (e.g., household and commercial, industrial, etc.) and, then, using a multinomial logit function[e] this total energy demand was partitioned into demands for particular fuels. The own and cross price elasticities were then calculated by comparing an original demand simulation with another demand simulation in which the appropriate price had been changed by 5 percent for all years. The percentage change in quantity demanded for each product was then divided by 5 percent to yield the desired own and cross price elasticities.

The regional model simply disaggregated these national forecasts to census regions using share equations. The estimated shares were functionally related to a variety of variables such as time, regional population, regional per capita income, and regional industrial activity, depending on the particular share being estimated. Regional prices were not included in the share equations. A more intrinsically appealing approach, one relying on regionally specified demand functions, was under development at the time, and has since become operational, but it had not been completed by the time the report was published.

---

[e]H. Theil, "A Multinomial Extension of the Linear Logit Model," *International Economic Review* X (October 1969): 251-59.

This demand modeling system was really one of the first attempts to model the demand for all energy products simultaneously, using prices as explanatory variables. Other previous studies had either modeled the entire system using various extrapolation techniques or had done an intensive demand estimation for specific energy sources. The former type of estimation had the prerequisite coverage, but it was neither historically accurate nor amenable to simulating the effects of price changes.[f] The latter type failed to deal completely with the question of fuel substitutability.

Success was not instantaneous for FEA either. The first complete runs in August of 1974 produced so many counterintuitive elasticities that the model had to be completely reestimated. Even though this reestimation resulted in marked improvements the results were never fully satisfying, particularly for natural gas, in time for the *Project Independence Report* issued in November 1974. Subsequent revisions have yielded a demand model that eliminates essentially all these problems.

In spite of their basic methodological soundness the uncertainty associated with these forecasts is quite high. The identification problem was never completely solved; the estimated price quantity relationships, to some unknown extent, represent the combined effects of supply and demand rather than the effects purely of demand. In addition, the regulation of natural gas since the 1960s made it impossible to obtain a recent demand curve for natural gas. Finally there was some question about the reversibility of the estimated demand curves. The estimates were derived from data characterized by falling relative energy prices. Their application to the problem at hand required simulating the model with rising energy prices.

The function of the linear programming supply model was to solve for the extraction, processing, conversion, distribution, and transportation activities needed to meet the predetermined demands as supplied by the demand model. Resource constraints could be explicitly incorporated, although, for reasons of time, and an inability on the part of the task forces to quantify these constraints in a model compatible format, they were never incorporated in the results that formed the basis for the published report.

The supply task forces were asked to approximate supply curves by a step function. In each case the task force was expected to produce a series of possible production levels and the minimum acceptable price that could call forth these production levels.[g]

The model was not dynamic. It was solved independently for each of three years, 1977, 1980, and 1985, for each scenario. Constraints on production from

---

[f]The limited accuracy of the extrapolation techniques was documented in Hans H. Landsberg, "Learning From the Past: RFF's 1960-70 Energy Projections," in Milton F. Searl, ed., *Energy Modeling: Art, Science, Practice* (Washington: Resources for the Future, 1973), pp. 416-36.

[g]The minimum acceptable price should be interpreted with care. It includes exploration and production costs plus-royalty and 10 percent after tax discounted cash flow. It does not include any costs of lease acquisition (e.g., bonus payments to the United States government for leasing an Outer Continental Shelf parcels).

reserves were built in to insure that the production estimates in the three years were compatible with each other and with the overall level of reserves.

The integration of the supply and demand models was accomplished by an iterative procedure involving price and quantity changes in the supply model. Each solution of the linear programming model yielded the minimum cost levels of each activity and also the shadow prices associated with each activity. When supply and demand prices were not equal, then the demand prices were adjusted in the direction of the supply prices. The resulting demand quantities were computed using the original demand price and quantity information plus all the computed own price and cross price elasticities. This method eliminated the need to run the demand model for every iteration. The convergence of this technique was never formally proven but, to date, it has not failed to converge. The major problem with the model, alluded to above in the demand section, was in handling natural gas. The initial estimates produced with the model gave a serious underestimate of the demand for natural gas, particularly at high oil prices. With the deadline getting ominously close the problem was traced back to two particular energy prices in the industrial sector, which were simulated to increase far beyond what previous econometric results would hold to be reasonable. The ad hoc adjustments to quote the *Project Independence Report* were:

In the industrial sector, the elasticity of each demand with respect to electricity prices was decreased by a factor of six and the elasticity of each demand with respect to natural gas price was decreased by a factor of four.[5]

A reestimation of the demand system based on regional (rather than national) prices has since eliminated this problem.

The main model in use for the economic impact analysis was an annual model, which linked an 89 sector input-output table with a long-term macro-econometric forecasting model.[h] This model was driven mainly by the energy price forecasts produced by the modeling system derived above. These energy prices were translated into product prices using the input-output table and markup equations. These product prices were then aggregated into a series of deflaters. The macro model would then compute personal consumption expenditures, disposable income, interest rates, a total for producer's durable equipment investment and the unemployment rate. The input-output model was then made to conform to these totals by scaling. This compatible model was then used to produce production, investment, and employment estimates by industry. These projections to 1985 were then used to drive an income distribution

[h]Documentation for the input-output model can be found in Clopper Almon, Jr. et al., *1985 Interindustry Forecasts of the American Economy* (Cambridge: Lexington Books, D.C. Heath and Co., 1974). The macroeconometric model was the standard Chase Econometric Associates, Inc., long-term model.

model, which assumed that constant shares of each of the functional types of income are received by each decile of the size distribution of income.[i] They were also used to drive a regional impact model, which used temporally changing regional shares of national employment levels in some 39 categories of employment at the national level to produce regional earnings estimates.[j]

The main deficiency with the economic impact results was a rather severe one: they were not completed until after the document was published! Since the economic impact forecasts were driven recursively by the energy sector forecasts, the economic impact forecasting could not take place until the energy sector results were available. The modeling system used to generate the energy sector results was continually being modified up until the last minute and satisfactory results were not available in time to analyze their economic impacts before the report was published. The main technical limitation of the macroeconomic impact forecasts was that they were inconsistent with the economic conditions used to initialize the demand model in some cases, and that the linkages between the energy model and the macroeconomic model were rather sparsely specified.

The energy-environmental tradeoff was considered an important policy question, which could benefit greatly from a clear analytical statement of the environmental impacts of alternative strategies. For many reasons the effort fell short of that. The original goal of including environmental constraints into the model proved impractical because of the nature of the environmental problem. For example, air pollutant standards, except for certain stationary sources and motor vehicles, are defined in terms of ambient concentrations rather than in terms of emission rates. In order to impose these ambient standards as constraints in the model the precise location of the facilities creating the pollutant would have to be known. In fact the model located facilities only in terms of a broad geographic area (e.g., census region).

The environmental analysis therefore was restricted to a quantification of the aggregate, rather than the local, impacts of alternative strategies for census regions. These impacts were derived simply by a matrix multiplication; the level of each generating activity (e.g., strip mining), as specified by the modeling system described above, was multiplied by a vector of constants to yield emissions of various pollutants and aggregate land requirements for each census region.[k] Because these aggregates bore no real relationship to actual environ-

---

[i]The methodology is described in J.M. Brazzel, "Effects of the 1973 Oil Embargo and Alternative Energy Strategies on the Distribution of Income" (Discussion Paper OEI-75-9, February 1975).

[j]The regional model is described in A. David Sandoval, "Multiregional Earnings Impact Model: Methodology" (Working Paper FEA-EAWP-75-5, May 1975).

[k]These factors can be found in Hittman Associates, Inc., "Environmental Impacts, Efficiency and Cost of Energy Supply and End Use" (Report to the Council on Environmental Quality, the National Science Foundation and the Environmental Protection Agency, September 1973).

mental objectives, this part of the analysis contributed very little to the formulation of energy policy.

Modeling for Policy

The energy modeling system described above was used extensively by the executive and legislative branches during their deliberations. Before that could happen, however, two problems had to be solved. The first problem concerned the fact that while the energy model described above is a long-term planning model, the politicians very much wanted to know what would happen in the next few years. Since the planning model provided only a 1977 forecast in the period to 1980, this did not provide enough information. As a result, for all petroleum forecasts the National Petroleum Demand and Supply model,[1] which had been developed as a quarterly forecasting model to assist in keeping track of impending developments, was modified to give three year forecasts. These forecasts were then combined with the 1980 and 1985 forecasts from the Project Independence model to characterize the effects of policies on petroleum demand and supply for the near and far term.[m]

The second major problem was how to condense the amount of information that could be generated by the modeling system in characterizing the consequences of any scenario generated during the policy formation process. Each scenario could lead to a variety of outcomes, depending on what happened to the world price of oil. While this uncertainty was a real and important dimension of the problem it created a problem in the policy process: it muddied the waters. As a result the decision was made by the policy makers that henceforth only one price scenario would be considered. That scenario would have the price of oil remain at $11 a barrel in real terms through 1977 and then fall to $7 and remain there through 1985.

This decision illustrates the distinct importance of controllability in the political use of analysis. This chosen price scenario, not surprisingly, was the one that made the President's program look better. With continued $11 prices the President's program was clearly too strong medicine.[n] At $7 prices it appeared much more reasonable. The choice of this particular price scenario did have a certain consistency with the goals of the program though, since one of the prime objectives of the Ford Energy Program was to force the world price for crude oil down to something like $7 a barrel.

With these problems solved the package of models was used extensively, both

[1]The documentation for this model can be found in Federal Energy Administration, Office of Quantitative Methods, *National Petroleum Product Supply and Demand: 1975* (Washington: U.S. Government Printing Office, 1975).

[m]As an interesting aside, the 1977 forecasts of the two models were quite close.

[n]The analytical support for this statement can be found in Chapter 9.

by the executive and legislative branches. Its main role in that process was to call the attention of the participants to the likely consequences of their actions on the major variables of interest. As a side benefit it diminished the use of much simpler, biased information. As shall be pointed out, such information was being used in both the executive and legislative branches prior to the use of this modeling system.

In the executive branch the contribution of the modeling system was to point out that the various components of the energy package as they had evolved during the Nixon administration and the early days of the Ford administration were not going to be as effective in reducing imports as expected. Whereas Secretary Simon had announced, before the analysis was completed, that coal production could reach 1.8 billion tons by 1985,[6] the analysis revealed that, when the limited substitution possibilities were taken into account, the maximum estimate was more like 918 million tons.[7] There was plenty of coal to produce, but the demand for it was limited. Similarly the analysis revealed that an accelerated program of Outer Continental Shelf leasing in the Atlantic, an administration objective, could be expected to add only 500,000 barrels a day.[8] It was also pointed out that new nuclear plant construction would not reduce petroleum imports very much. Rather, it was more likely that new nuclear plants would replace new coal-fired plants.[9] Thus, the previous estimates of the relationship between these policies and imports, which simply added up the separate contribution of each, overstated the reduction in imports because of double accounting. The reaction to this revelation by the central policy figures was not to put less emphasis on these programs, but rather to add other elements to the package to reduce projected imports to their previously expected levels.

One example of this was the addition of the import fees to the President's energy program. When it became very clear that the 1 million barrels a day target for import reduction in 1975, set by the President on October 8, 1974, would never be reached by the voluntary means proposed, the import fees were added.[o] The model was used both to derive the level of these fees, which would meet the 1 million barrels goal, and to trace out its implications for the longer run.

The use of this modeling system also served to identify likely long-run implications and some potential problem areas. More intensive studies of these problem areas were then initiated. For example, special studies on the natural gas situation with and without deregulation, on the problems of electric utility financing, and on siting future power plants were undertaken when the initial forecasts suggested that these were pivotal issues.

The models themselves are continually undergoing changes to eliminate the bugs that were discovered during the writing of the *Project Independence Report.* There is currently in the works a plan to redo the entire report using

[o]The politics behind the construction of the package is described in some detail in Chapter 7.

these refined models and, if all goes according to plan, it will be published before this book.

The models were also used by Congress. Certain congressional committees such as the House Ways and Means Committee were given access to the models and staff of the Federal Energy Administration and the Treasury Department. In this way these committees could, and did, examine the implications of a number of proposals on energy consumption, petroleum imports, revenues, and expenditures, and on various measures of impact on the economy. All of these were provided shortly after the committee scenarios were developed, in some cases less than 24 hours later. This interaction was handled on a staff to staff basis rather than through channels. This arrangement met the controllability and timeliness criteria for the committees at some sacrifice of executive branch controllability.

This relationship was exploited heavily by both committees. Because of the differential speed with which the two committees were working, the analytical support was concentrated first on the Ways and Means Committee. Then, when that bill had passed the House, the support was shifted to the Energy and Power Subcommittee of the House Interstate and Foreign Commerce Committee, which was considering the various decontrol and deregulation measures. How important this information was to the congressional decision-making process is unclear, but the fact that the various members of the committees had access to it was significant in its own right.

The relationship between the congressional and administrative staffs seemed to grow from one of disdain and outright suspicion to a healthy one of cautious mutual respect. Nothing was ever accepted at face value. Assumptions were always questioned and further evidence sought on many occasions. While resolution of the differences was not always achieved, the result of this interaction was certainly a more careful consideration of the impacts of the proposed policies than would otherwise have been the case if no modeling system had been available.

Up to this point we have discussed the use of the modeling system in the policy process. The other tangible output, the *Project Independence Report*, was also used by the process, but with noticeably less success. This point can be illustrated by relating two incidents that occurred during the early deliberations.

The staff who derived the background materials for the Democratic leadership energy plan developed their estimates from the *Project Independence Report*.[P] Numbers were drawn from various places in the report and simply added together. These impacts were subsequently found, when the FEA staff got together with the congressional staff, to include some double accounting. It was also discovered that some of the numbers from the report, which were used, were generated using assumptions that were inconsistent with the Democratic

---

[P]These estimates can be found in Gene Kinney, "Demos Reveal Alternative Energy Plan," *Oil and Gas Journal*, 10 March, 1975, pp. 26-28.

plan. These impacts were not an important part of the policy process, but it does illustrate the principle that reports rarely satisfy the needs of all readers. You cannot address "What if . . . ?" questions to a report.

The second incident proved quite embarrassing to the administration.[q] On March 18, 1975, when Administrator Zarb was testifying before the Senate Commerce Committee, Senator Adlai Stevenson, III (D-Ill.), waving a copy of the *Project Independence Report* in the air, asked why the FEA insisted that higher natural gas prices would lead to increased supplies, when its own report indicated that that was not so?[r] The table in question did indeed show that, but its inclusion in the report was an editorial mistake. The model used to generate those estimates only had valid input data up to 80 cents per mcf. The model generated estimates for higher prices as part of the standard procedure, but the outputs for these higher prices were meaningless. Unknowingly, someone else took this output, created the table in question, and in the rush to publication it was published by mistake.

In retrospect it appears that the Project Independence experience is not compatible with the pessimistic hypothesis that the inability to estimate the consequences of alternative packages would seriously retard the responsiveness of the policy process. Many of the natural forces impeding the use of analysis in the decision-making process were able to be overcome.[s] In part this was due to the ability of the particular persons involved to provide an effective channel into the process and in part it was due to the design characteristics of the modeling system. The shortness of the development period imposed very real costs on the analysis; these have been detailed above. Yet, the most serious of these analytical shortcomings probably had less adverse affect on policy in the long run than the very positive benefit derived from the early availability of the system, namely, that the channels have now been established and procedures worked out for integrating the latest analysis into the process. While the ultimate success of this integration remains to be seen, it does provide a basis for optimism on the prospects of infusing more analytical information on complex subjects into the policy process.

---

[q]An official version of this incident can be found in U.S. Congress, Senate, *Natural Gas Production and Conservation Act of 1975, Hearings before the Committee on Commerce, United States Senate*, 94th Cong., 1st sess., 1975, p. 118.

[r]The data in question can be found in Federal Energy Administration, *Project Independence Report*, p. 93.

[s]Most of the tension between politics and analysis seemed to be handled by the technique of selective omission. Rather than sending the analysis back asking that "better" numbers be produced, those estimates that would have tended to bolster the congressional Democratic position were never permitted to see the light of day, although they did circulate widely within the executive branch. This, for example, was the case with the economic impact statement that was accomplished to estimate the effects of implementing the president's energy program on the economy. Accomplished soon after the congressional debate had begun, it was circulated within the executive branch only. Even this technique was used sparingly, however, and most FEA estimates were documented and made available to any interested party in a series of technical papers.

Notes

1. Edward Cowan, "Zarb is Praised as Man in Middle on Oil Price Dispute," *The New York Times*, 23 July 1975.

2. Roland Evans and Robert Novak, "Zarb Makes His Mark on the Hill," *The Washington Post*, 12 April 1975.

3. See Kenneth C. Hoffman, "A Unified Framework for Energy System Planning"; D.E. Deonigi and R.L. Engel, "Linear Programming in Energy Modeling"; and James M. Griffin, "Suggested Roles for Econometric and Process Analysis in Long-Term Energy Modeling." All are found in Milton Searl, ed., *Energy Modeling: Art, Science, Practice* (Washington: Resources for the Future, 1973), pp. 103-85.

4. Policy Studies Group, *The FEA Project Independence Report: An Analytical Assessment and Evaluation* (Cambridge: MIT Energy Lab, 1975); and Battelle Memorial Institute, *Review of the Project Independence Report* (Columbus, Ohio: Battelle Memorial Institute, 1975).

5. Federal Energy Administration, *Project Independence Report*, (Washington: U.S. Government Printing Office, 1974), appendix, p. 87.

6. Congressional Quarterly, *Continuing Energy Crisis in America* (Washington: Congressional Quarterly, 1975), p. 3.

7. Federal Energy Administration, *Project Independence Report*, appendix, p. 47.

8. Federal Energy Administration, *Project Independence Report*, p. 83.

9. Ibid., p. 8.

Part III:
The Policy Process

7 Formulating an Administration Position

Having examined, in the light of the Project Independence experience, the consistency of the evidence with the hypothesis that the absence of information would impede the responsiveness of the policy process, it is now time to turn to an examination of the remaining two hypothesized constraints. These hypotheses suggest a relationship between the policy process itself and the nature of the outcomes produced by that process. The first states, in brief, that the internal organizational structures of both Congress and the executive branch are inappropriate for developing a comprehensive and coherent energy policy and furthermore that they are resistant to change. The second hypothesis states that the issues are too politically volatile and too divisive to be resolved in a participatory democracy. In Part III we assess the consistency of the evidence with these two general hypotheses by examining and evaluating the policy process as it operated during the first two years of Project Independence.

There is a popular notion that the executive branch is ideally suited for policy formulation. Unencumbered by the necessity to form coalitions of politicians seeking to satisfy quite different constituencies, supported by a large staff, and unified by a hierarchical decision-making structure, the executive branch is viewed as the epitome of an efficient, centralized decision-making structure. While, in comparison to Congress, there is a great deal of truth in this characterization, it oversimplifies greatly the actual process.

The executive branch is not a monolithic organization with the President in complete control. It has multiple centers of power with which the President must deal, and not always as the ultimate authority.[1] A shared ideology is not a sufficiently strong commonality to guarantee that the various key members of the administration will agree on particular policy options. As a result compromises are necessary, which, when unaccompanied by a description of the historical struggle that preceded them, will appear confusing to the general public.[a]

The process by which the Ford administration formulated its package of energy proposals is a case in point. The key questions about this program are: What factors influenced the choice of policy instruments and the degree to which they would be applied? What factors led to the politically volatile decision to implement the higher price portions of the program immediately rather than phasing them in over some longer period? Why, after a long concentration on

---

[a]This point is particularly well developed in Morton Halperin, *Bureaucratic Politics and Foreign Policy* (Washington: Brookings Institution, 1974). See especially chapters 1 and 16.

accelerating domestic supplies as the basis for policy in the Nixon and Ford administrations, did President Ford come out strongly for stringent conservation measures? The quest for answers to these questions begins with a discussion of energy policy during the Nixon administration.

Energy Policy in the Nixon Administration

The Nixon administration placed its imprint on energy policy in a number of ways, not all of which were intentional. The main intentional accomplishments of the Nixon administration were a restructuring and improvement in energy decision-making procedures in the executive branch, the decision to give a green light to the construction of the Alaskan pipeline, and the initiation of the effort to apply comprehensive planning models to the policy process. The main unintentional effect was the influence that the Watergate crisis had on energy policy, both by providing a diversion for public opinion, which allowed time for the analysis to take place, and by being an important causal factor in changing the composition of Congress. This latter point is explored in more depth in Chapter 8.

The first Nixon initiative on energy matters was contained in a message to Congress on June 4, 1971. This was a historically significant document in that it marked the first time in United States history that a presidential message to Congress had focused exclusively on energy. This message contained several proposals to expand the domestic supplies of energy: development of the nuclear breeder reactor, the acceleration of oil and gas leasing on the Outer Continental Shelf, and an expansion of the domestic capacity to enrich uranium, the fuel to be used for nuclear reactors. The message also contained a request for legislative approval for the creation of a new Department of Natural Resources.

The Nixon program rested heavily on the development of the nuclear breeder reactor as the best domestic source for clean, dependable energy. The energy problem was seen as being concerned with import vulnerability, environmental preservation, and resource exhaustion. The specter of import vulnerability had been raised by a Cabinet Task Force a year earlier.[2] Public support for environmental preservation was strong. This public support, for example, had resulted in the National Environmental Policy Act, which was signed by President Nixon on January 1, 1970.[b] This act, among other things, required environmental impact statements for any government action significantly affecting the quality of the human environment. The perennial concern over resource exhaustion was precipitated by the lack of agreement on the size of the United States petroleum reserves. If the smallest estimates were correct, then, unless

[b]A description of this vitally important act and its far reaching consequences can be found in Council on Environmental Quality, *Environmental Quality: The Third Annual Report* (Washington: U.S. Government Printing Office, 1972), pp. 221-67.

some acceptable substitute were available, increasing the domestic production of oil and gas would simply hasten the time when domestic resources were depleted.

President Nixon focused his attention on the nuclear breeder reactor because it provided what Nordhaus has subsequently called a "backstop technology."[3] A *backstop technology* is one that provides an acceptable substitute for the scarce resource and that rests on a very abundant resource base. Its role is to make domestic resource depletion a much less important concern by providing the capability to substitute an abundant resource for the scarce one. The nonbreeder nuclear reactor did not provide that function because it depended on U^{235}, a fuel in rather limited supply. The breeder reactor, on the other hand, would use U^{238} as its basic fuel and this fuel is 140 times more abundant.[4]

This program, however, did not receive unanimous acclaim. The Interior Department plan to accelerate the sale of oil and gas leases on the Outer Continental Shelf ran into a suit by the Natural Resources Defense Council. In *Natural Resources Defense Council v. Morton* the Court of Appeals for the District of Columbia held that the Department of Interior had failed to comply fully with the National Environmental Policy Act in drawing up the associated environmental impact statement for its Outer Continental Shelf leasing program.[5] The result was an inevitable delay in implementing the plan, although the time was used to conduct a more thorough assessment.[6]

The plan to accelerate the development of the nuclear breeder reactor was more successful. By fiscal year 1973 over one-half of all federal research and development funds were allocated toward the development of nuclear fission power and about two-thirds of that went directly toward research on the breeder reactor.[7] Following the Nixon speech, however, the critics of nuclear power became more vocal. They pointed out that although nuclear power was clean from an air pollution point of view, it was potentially very dangerous.[c] The dangers arose both from the actual operation of the plant and in the handling and disposal of the highly radioactive spent waste. In this kind of volatile political climate neither the President nor Congress forced the initiative on energy policy. By early 1973, however, it became clear that because of rising demand, falling domestic production, and quotas on petroleum imports, shortages of petroleum products would occur later in that year in the absence of some positive action to alleviate the situation.

On April 18, 1973 the President sent a new energy message to Congress. This message announced the termination of the 14-year-old oil import quota system. It also contained a renewed emphasis on the offshore exploration for oil and gas and a partial decontrolling of natural gas prices. In addition it proposed the construction of deep-water ports to handle huge oil tankers and a $130 million increase in funds for research and development of future energy sources. The

---

[c]The nature of this debate and some references can be found in Gerald Garvey, *Energy, Ecology, Economy* (New York: W.W. Norton & Co. 1972), chapter 7.

overall impact of the program was to rely on increased imports in the short run and increased domestic supplies in the long run.

In October of that year the Arab oil embargo imposed a new urgency on the problem. As described in Chapter 3, public opinion was aroused by the embargo and it concentrated the blame on the government, not the Arab nations. It became clear that a new bold move was needed from the President to counteract this negative public opinion. He responded with what later became known as the "Project Independence" speech on November 7, 1973.

The Project Independence speech was one of two major cases in the recent history of energy policy in which the political environment demanded answers, bold answers, before the alternatives had been clearly thought out and weighed.[d] Nixon chose to respond to this situation by setting up a massive effort to sift through the alternatives and come up with a comprehensive plan. Realizing, however, that this strategy would probably fall short of satisfying public demands for swift and decisive action, he set a tough goal for this effort:

Let us set up as our national goal . . . that by the end of the decade we will have developed the potential to meet our own energy needs without depending on any foreign sources.[8]

That this goal was highly unreasonable, and probably infeasible, was a widely held belief among energy experts at the time. Luckily, the goal was not binding on the Project Independence analysis because, before the analysis was completed, Nixon resigned. President Ford, having no personal stake in this goal, was quick to accept the advice of his advisors to quietly bury the notion of self-sufficiency by 1980.

This speech points out a fundamental dilemma in long-range planning in a democracy. When the issues are not in the foreground of public opinion, more time is available for assessing choices and making rational decisions, but there is less pressure on Congress to enact the required legislation. The result may be no firm policy at all. On the other hand when public opinion is aroused, the chances for action are enhanced, but the immense pressure for immediate action forces decisions before a comprehensive package can be developed.

The lasting positive contributions of the Nixon administration were the direct result of the pressure put on Congress by the Arab oil embargo. On November 13, 1973, less than one month after the initiation of the embargo, Congress ended several years of debate and authorized the immediate construction of the Trans-Alaska pipeline. This law opened the vast Alaskan reserves to the United States market. When completed, it would lower import dependency by 1.5 million barrels a day.

The other main contribution of the Nixon administration was in the area of

---

[d]The other major case was President Ford's speech on October 8, 1974, which committed the nation to a million barrels a day import reduction. This case is described below.

reorganizing the responsibility for energy matters. The existing jurisdictional divisions were based on a multitude of unrelated decisions. The Federal Power Commission had control over natural gas, the Atomic Energy Commission had control over nuclear power and the Department of Interior had control over coal and the Outer Continental Shelf. There was a great deal of suspicion in Washington that each agency had, by tradition, become a partisan supporter of its particular fuel to the detriment of a more comprehensive view. This hodgepodge of scattered jurisdictions with no overall coordinating structure was clearly an impediment to effective comprehensive policy making and was widely recognized as such.[e]

The President's initial attempts at executive reorganization were rebuffed by Congress. His request for congressional authorization for a Department of Energy and Natural Resources, first conveyed to Congress in 1971 and requested again on April 18, 1973, never progressed beyond the hearings stage.[9] In response to the embargo he created the Federal Energy Office on December 19, 1973 by an executive order to manage the embargo and the development of an energy plan. He also created the Energy Resources Council to coordinate top level energy policy making. On May 2, 1974 Congress authorized for a period of two years the Federal Energy Administration, which replaced the Federal Energy Office. Nixon had also requested in a June 29, 1973 speech a new independent agency to handle all energy research and development, to replace the separate research and development operations that were scattered throughout the executive branch. This recommendation, with extensive modifications, became law in October 1974 after Gerald Ford had assumed the presidency.

Prelude to the Ford Energy Program

Every new president gets a honeymoon period in which the political climate is tranquil and benign. Although everyone working on the Project Independence analysis fervently hoped that this grace period would last until the *Project Independence Report* was in, it was not in the cards. The honeymoon was terminated by a series of events that soured public opinion. President Ford gave an executive pardon to former President Nixon, which aroused a good deal of controversy. Meanwhile the economy showed serious signs of weakening. On September 12 the Labor Department announced that the second largest monthly rise in the seasonally adjusted wholesale price index since 1946 had occurred in August. On September 18 the Commerce Department reported that the United States balance of payments showed a $2.7 billion deficit during the second quarter of 1974. In both cases high oil prices were given as a major contributing

---

[e]A somewhat dated, but largely still relevant, discussion of the problems associated with policy making in this kind of environment can be found in "The Case for a Department of Natural Resources," *Natural Resources Journal* I (November 1961): 197-206.

factor. The pressure for a firm energy policy mounted. In summarizing the views of the Economic Summit Conference, convened by the President to examine alternatives for the economy, one observer cited the virtual unanimous agreement on the need for a firm energy policy as soon as possible.[10]

While this pressure was mounting developments were taking place within the administration that would have an important effect on policy. The first and most important of these developments was that Secretary of State Henry Kissinger had concluded that United States energy policy must include a sharp, immediate curtailment in imports. He was an important figure in the energy policy deliberations because he had access to the President and because he had staked out a well-articulated claim over energy policy as a vital link in his overall foreign policy strategy.

His access to the President on energy matters was achieved largely because he had maintained a consistently high degree of power and prestige during the turbulent later years of the Nixon administration. While the various energy advisors to the President generally had short tenures in office, Kissinger was continually in charge of foreign policy.

His response to the energy problem was shaped by his conception of the long-term requirements of American foreign policy. This conception, formulated before he became a member of the administration, revolved around the perceived need for the evolution of a new concept of order in the international system to replace the increasingly disjointed post-World War II order.[f] Kissinger's efforts during the Nixon administration, most notably the Soviet-American detente and the rapprochement with the People's Republic of China, were designed to insure that the new world order would not take a form destructive to America's interest and world leadership. In early 1973, after having made much progress in these areas and after having apparently ended the Viet-Nam War, Kissinger attempted to turn his attention towards the allies. The post-World War II Atlantic Alliance, the cornerstone of American foreign policy for a quarter century, was in serious trouble, a fact Kissinger had long emphasized.[11] The reduction in the Soviet military threat, the growth of European economic strength and political independence, America's tendency to act unilaterally without consulting allies, and its preoccupation with Asia had all resulted in divisions within the alliance. As Kissinger later put it, "The military organization and the political and economic organization had grown out of phase with each other."[12]

Such a situation was not conducive to the evolution of a new world order. Kissinger's initial attempt to reform and revitalize the Atlantic alliance was heralded by his April 1973 "Year of Europe" speech, in which he singled out the

---

[f]A concise analysis of the need for a new concept of order and the barriers inhibiting its creation is found in his essay "Central Issues of American Foreign Policy," which can be found in Henry Kissinger, *American Foreign Policy* (New York: W.W. Norton and Co., 1969), pp. 51-98.

need for an insured supply of energy as a new reality requiring cooperative action.[13] The European response was not enthusiastic and the proposal lay dormant for a number of months. However, the Organization of Arab Petroleum Exporting Countries (OAPEC) embargo and the accompanying rise in the price of imported crude oil a few months later gave a new impetus to Kissinger's proposals.

These actions by the oil exporting nations provided both a threat and a promise to Kissinger's hopes concerning a new framework for world relations. The vast flows of dollars to the oil exporting countries created a powerful new force in the world, which would complicate the relationships with the allies. There would be a temptation for each nation to embark on a highly competitive race with each other to sign long-term bilateral agreements with the producing nations to insure themselves of a steady supply of oil. As subsequently became clear, this was a realistic concern.[g] The problem with the bilateral approach lay not only with the fact that it would make a cooperative approach more difficult, but also that it would tend to institutionalize and perpetuate high oil prices. Kissinger was clearly worried that continued high prices could lead to a worldwide depression, economic retaliation, and political upheaval in the West, which could diminish the strength of the Western Alliance vis-à-vis the Communist nations.[h] The Communist nations, being self-sufficient in energy, could be expected to remain relatively immune to the oil upheaval.

There were other threats as well. The independence of United States foreign policy in the Middle East was at stake since the oil exporting nations had demonstrated both the resolve and the capability to use oil as a political weapon. The stability of the Middle East was also threatened to the extent that the newfound wealth was used to finance military weapons to be used against Israel.

The promise behind the threat was that since the threat was shared by the members of the target alliance this could serve as a bond for joint action. The objective of this joint action, in the Kissinger strategy, was to be a lowering of the world price for crude oil. The means of achieving this goal was to be negotiation,[14] a tactic Kissinger had found quite successful in resolving other disputes.

Kissinger's plan called for the development of a unified consuming nation

---

[g]France, traditionally the maverick in the Atlantic Alliance, led the other European states in seeking such agreements. By January 1974 France had apparently agreed with Saudi Arabia to exchange military hardware, petrochemical factories and refineries for a steady supply of crude oil over a 20-year period. See *New Republic*, 23 February 1973, p. 10.

However France was far from alone. By that time the United Kingdom, West Germany, and Belgium, as well as Japan, were all reportedly seeking bilateral agreements. *Business Week*, 19 January 1974, p. 19.

[h]For two examples of this theme in his speeches, see Henry Kissinger, "Major Oil-Consuming Countries Meet in Washington, D.C. to Discuss the Energy Problem: Statement by Secretary Kissinger," *Department of State Bulletin* LXX (January-June 1974): 201, and "The Energy Crisis: Strategy For Cooperative Action," *Department of State Bulletin* LXXI (July-December 1974): 750.

negotiating strategy prior to meeting with the oil producers. The core of this policy would be a collective reduction in the demand for OPEC (Organization of Petroleum Exporting Countries) oil. This would serve the twin purposes of demonstrating the resolve of the consuming group to OPEC while inflicting a revenue loss on the producers. Since it was believed by Kissinger that OPEC would be unable to curtail production sufficiently to maintain these high prices in the face of this diminished demand, they would be willing to come to the conference table.

The Washington Energy Conference, made up of delegates from the oil-consuming states, met in February 1974 to consider the Kissinger proposals. Of the areas of needed cooperation he stressed the need for an immediate reduction in demand.[15] He also proposed to share American oil in the event of another embargo.

In spite of this offer the allies approached the proposal with great caution. Europe and Japan, far more vulnerable to OPEC than the United States, were loath to antagonize the producers by joining what was essentially a confrontation group. Furthermore, many Europeans apparently saw the proposal as a vehicle for a new American subordination of its allies, by grace of its large domestic reserves of energy.[16] European fears were reinforced by Federal Energy Office (FEO) Administrator Simon's refusal to commit the United States to gasoline rationing, even if such a policy were needed to implement a plan for sharing the scarce oil among the consuming nations, and by the contradiction implied between President Nixon's Project Independence and Kissinger's theme of interdependence. Prior to the conference, French Foreign Minister Michael Jobert had obtained a common market vote to reject Kissinger's plan for a permanent consumer organization.[i]

The conference did achieve some of Kissinger's objectives, however. A temporary high level group to consult on energy matters, the Energy Coordinating Group, was formed. In addition, the following fall he was able to overcome French resistance and the International Energy Agency (IEA) was formed. This agency was set up to coordinate joint conservation and research efforts as well as to manage the emergency sharing system known as the International Energy Program. These successes, however, could not camouflage the demonstrated European lack of faith in United States intentions and strategies. The French, for example, refused to join the IEA at all.

While Kissinger was experiencing difficulties abroad, he was enjoying somewhat more success at home, at least with the President. His impact on the President became clear when in September 1974, in a speech to a World Energy Conference, the President echoed Kissinger's firmness on the necessity for the world price of oil to fall and even added a hint of possible military action.[17] The speech was an attempt to persuade Europe and Japan that the foreign policy

---

[i]See, for example, the French reaction in *New Republic*, 23 February 1974, p. 10.

aspirations of interdependence were completely compatible with the domestic pursuit of Project Independence.

The allies remained unconvinced. In answer to Kissinger's proposal concerning general consumption reduction by the oil consuming nations the British Chancellor of the Exchequer Dennis Healy tersely replied that because the United States wasted so much more energy than Europe it should bear the brunt of further reductions.[18]

At that time there was some optimism that a price cut could be forthcoming if the consuming nations appeared determined in their efforts to reduce imports. On his way to the same World Energy Conference addressed by President Ford, Saudi Arabian Oil Minister Ahmed Zaki Yamani indicated that world oil prices were perhaps $2/bbl too high and should come down.[19] This appearance of support for lower prices within OPEC itself probably helped to solidify Kissinger's position within the administration that the time appeared right for a bold move.

The second development that occurred was an increasing mood within the White House that conservation should be a part of the strategy. This was due both to Kissinger's push for a solid demonstration of willingness to sacrifice (which could not be accomplished by simply switching from foreign to domestic energy sources because most other nations did not have this capability) and to the realization that the long lead times for increasing domestic supplies meant that, in the short run, almost all import reduction would have to come from conservation.

In preparation for a major address on energy and the economy on October 8, 1974, the Federal Energy Administration submitted a list of recommended conservation measures to the President, including a gasoline tax. Higher gasoline taxes were apparently widely supported within the administration,[20] but their chief supporter, John Sawhill, was not, as became clear on October 29, when he was forced to resign. Because of this the Federal Energy Administration, and the analytical apparatus that it controlled, played essentially no role in the President's October 8 speech.[j]

In that speech the President sided with Kissinger in committing his administration to a 1 million-barrel-a-day import reduction by the end of 1975, but it was somewhat weak support because he rejected gasoline taxes as a vehicle for achieving this goal. Bowing to the political requirements of the upcoming election in November, Ford suggested that the reduction would be accomplished by voluntary measures. The speech was important because it committed the administration to a specific goal before the consequences of that goal had been carefully thought out. This goal was subsequently to become an important constraint in assembling the Ford energy program.

---

[j]In fact Interior Secretary Morton was quoted at the time as derogatorily referring to the analysis as "a lot of fancy footwork with computers." See Joel Haveman and James G. Phillips, "Energy Report/Independence Blueprint Weighs Various Options," *National Journal Reports*, 2 November 1974, p. 1635.

The Ford Energy Program

On December 14, a group of energy advisors met at Camp David to prepare a report listing his options to President Ford. The principle figures in attendance were Interior Secretary Rogers C.B. Morton; Alan Greenspan, chairman of the Council of Economic Advisors; John Hill of the Office of Manpower and Budget; William Seidman, a presidential economic advisor; Thomas O. Enders, the assistant secretary of state for business and economic affairs; Transportation Secretary Claude Brinegar; FEA Administrator Frank Zarb; Assistant Treasury Secretary Parsky; and Eric Zausner.[k]

The deliberations did not start from scratch. There was a history, in the Nixon and Ford administrations, of policy pronouncements on energy. It should not be surprising that these policies survived not only the Camp David meetings, but the subsequent meetings at Vail, Colorado as well, where the final economic and energy program was assembled. Policies falling into this category were the acceleration of the coal leasing program, acceleration of the Outer Continental Shelf leasing program, the deregulation of natural gas, and the more rapid development of nuclear power.

The Federal Energy Administration (FEA) modeling system was by this time fully integrated into the policy process. In preparation for the Camp David meeting detailed briefing books were prepared outlining a range of options and their consequences. The Project Independence analysis was the major source of information consulted in preparing these briefing books. In retrospect the role it played was to point out that this historical policy package was not sufficient to meet the administration's goals either in the immediate future or in the long run, if the world price were to fall. The President's October 8 speech had set a tough goal for import reduction and it was clear it would never be achieved by the voluntary measures originally proposed.

Thus, the President had to decide whether or not the goal would stand in light of the more stringent measures needed to meet it. If it was to stand, the policy instruments to achieve the goal had to be selected. The goal stood. One of the rejected options was a more moderate goal, which would have allowed phasing in the higher prices resulting from the deregulation of natural gas and the decontrolling of domestic oil prices over a period of four years, thus moderating the adverse effects on inflation and unemployment. This option was specifically rejected, apparently because it would not make a forceful enough demonstration of United States leadership to Europe and Japan.[21]

The choice of instruments was made largely on political grounds. The gasoline tax was considered, but rejected. Because the President had come out so strongly against a gasoline tax in his October 9 press conference,[22] changing his mind

---

[k]The list of attendees was obtained from Bruce Pasternack, currently the deputy assistant administrator for policy at the Federal Energy Administration, who attended the Camp David meetings.

would have required a politically unpalatable position reversal. The most compelling argument for rejecting the gasoline tax, however, was that its imposition required additional legislative authority while import fees or quotas could, it was thought, be unilaterally imposed by the President.

The President was authorized by the Trade Expansion Act of 1962 to set import fees or quotas upon determining that this action was necessary to protect the national security of the United States.[1] The ability to unilaterally impose these fees was an important political consideration because, it was felt, this action would serve the twin purposes of letting the country know he was asserting decisive leadership and of prodding a reluctant Congress into action. If all went according to plan these fees would be replaced by a congressionally authorized tariff. The import fees would be purely a stopgap measure.[m]

The selection of an import fee rather than a quota was largely an ideological preference. Secretary of State Kissinger had originally backed the quota, but Simon and others preferred the use of import fees. Because import fees required less bureaucratic control, they would be more consistent with another main objective of the Ford administration, getting the government out of the marketplace.

Simon and Kissinger also clashed over the desirability of a guaranteed floor price for petroleum. The secretary of State backed the idea, which was first proposed publicly in November 1974 by Kissinger's assistant secretary for economic and business affairs, Thomas O. Enders. After a great deal of discussion within the administration, Kissinger was able to announce vague American support for the concept of a price floor on February 3, 1975. However, Simon and other opponents of the idea had succeeded in blocking full administration support, leaving the idea as only a possible option.

The issue in large part reflected the differences in approach towards economic issues between the two men. As outlined in February[23] the Kissinger price-floor plan was designed to guarantee that the price of petroleum would fall no lower than a certain level. The price floor was an essential part of the Kissinger negotiation package. It represented the carrot, whereas the demand reductions represented the stick. In combination, in Kissinger's view, they would provide a

---

[l]The description of the use of these fees and the court rulings on their legality can be found in *National Journal Reports*, 1 March 1975, p. 321 and *Facts on File*, 23 August 1975, p. 607.

[m]All did not go according to plan. The constitutionality of the fees was immediately challenged in the courts. Following a favorable ruling for the President on February 21, 1975, which denied a motion for injunction, a U.S. Court of Appeals ruled on August 11, 1975 that the President had exceeded his authority. Because this decision was immediately appealed to the Supreme Court, no attempt was made to stop collecting the import fees while the appeal was pending, however. In its decision the court specifically held that the normal system of checks and balances could not be repealed even in the face of an emergency. Thus, the judicial system responded to the issue of institutional responsiveness by disallowing the unilateral move of the President, but it took so long to do so that the import fees were actually imposed, apparently in violation of the law, for a number of months.

powerful stimulus for negotiating the world price of crude oil downward. The demand reductions would introduce an uncertainty among OPEC members over the stability of their cartel. The floor price would guarantee them continued prices that would exceed the costs of production. Therefore, in return for a reduction in the price from current levels, they would be able to develop their economies unthreatened by the prospect that the price for their oil would ever again fall to equal the cost of production. Furthermore, in the Kissinger view, this negotiation would be in the best interests of the consuming nations because it would result in a lower price while protecting the development of alternative energy sources to reduce their dependence on OPEC in the long run.

In opposing the Kissinger strategy, Simon did not deny the necessity of protecting alternative energy source investments. This was an objective shared by practically all members of the administration.[24] Simon did, however, oppose Kissinger's choice of mechanisms. Because the imposition of an artificial price floor ran counter to Simon's firm ideology supporting the maintenance of a free market, he backed a proposal for a fixed tariff, to be added on to any OPEC price. This, he reasoned, would protect and encourage investments, while allowing a break in OPEC prices to be reflected in lower costs of imported oil. The Kissinger strategy, he argued, would institutionalize higher than necessary oil prices at the expense of consumers.[n] Compared to a tariff, which would protect energy sources while channeling the revenue back into the domestic economy, the price floor would accomplish the same objective by channeling the revenue to OPEC. This, argued Simon, would be a little like becoming a nonresident OPEC member in which you pay a fee, but are denied the benefits of membership.[25]

The public clash between these two central participants began with the Enders speech, given at Yale on November 25, 1974. This speech represented the first explicit plan for an oil price floor, although the concept had been introduced earlier.[o] In that speech Enders apparently called for guaranteed prices at "current levels," which was widely interpreted to mean $11 a barrel.[26]

Simon's reaction was swift and hostile. Kissinger immediately retrenched somewhat by classifying Ender's statement as "unofficial" and restricting its distribution.[p] On December 13 Kissinger reiterated his support for lower prices

[n]Simon may have received important support from Federal Reserve Board Chairman Arthur F. Burns, whose views mirrored those of the Treasury secretary. Burns considered Kissinger's idea to be an "outrage," a "blunder," and a "virtual surrender to OPEC." See Leonard Silk, "Market vs. State," *The New York Times*, 19 February 1975, p. 47. Although not officially involved in energy policy, Burns was believed by many to have strongly influenced it in the past. See *Business Week*, 28 September 1974, pp. 30-31.

[o]For an example, see Henry Kissinger, "The Energy Crisis: Strategy for Cooperative Action," *Department of State Bulletin* LXXI (July-December 1974): 749.

[p]The speech is not to be found in the *Department of State Bulletin*. Furthermore, Yale University, the State Department, and Enders' office will not give out copies. See Eliot Marshall, "Oil Double-Talk," *New Republic*, 5 and 12 July, 1975, p. 11.

and described the Enders proposal as one that was simply under some study.[27] There can be little doubt, however, that Kissinger knew of and, indeed, had approved of the Enders speech before delivery.

At the Vail meetings, which Simon attended but Kissinger did not, it was decided that the President would propose legislation that would authorize, but not require, the President to set a guaranteed price. After Vail, Kissinger apparently prevailed on the President to change the wording to require the establishment of a minimum floor price.[28] The final compromise yielded proposed legislation that required the President to set some form of price floor by a tariff or other means, but provided a large number of loopholes should he decide that such a step would not be in the national interest.[29]

The other parts of the energy policy package were internally less controversial. The main political objections to the program, as outlined above, could be anticipated to be that it was unduly hard on the poor and that it would create windfall profits for the oil and natural gas companies. To compensate for the oil company profits resulting from decontrol of crude oil prices a windfall profits tax was included. This tax would phase out over time. To prevent the tariff on oil imports from providing an incentive to energy consumers to substitute the increasingly scarce natural gas for the imported oil an excise tax of 37 cents per thousand cubic feet of natural gas was suggested. This level of tax was chosen because it is equivalent to the $2 tariff on petroleum products on a BTU basis. To protect the poor, $2 billion of the collected energy taxes would be set aside for dollar rebates to nontaxpayers and low-income taxpayers.

The remaining elements of the package, proposals for relieving the financial problems of the electric utilities industries, the creation of a strategic petroleum reserve, and a number of conservation actions, came directly from the Project Independence analysis. The analysis of the financial outlook for utilities had produced a gloomy forecast,[30] an outlook shared by some academic observers.[31] The result was the inclusion in the policy package of a number of remedial actions: an increase in the investment tax credit for utilities, a number of regulatory reforms (e.g., automatic fuel adjustment pass-throughs and allowing pollution control expenditures to be counted in the rate base), and an administrative charge to the Energy Resources Council to conduct an intensive study of the problem and make recommendations. This charge reflected the degree of uncertainty about the effects of higher electricity prices on electricity demand growth, a key determinant of the seriousness of the problem.

The strategic reserve had been strongly supported in the *Project Independence Report* as a very cost-effective way to reduce import vulnerability. In fact, it appeared cheaper than a good number of other domestic strategies. The decision process, however, never performed this trade-off analysis. Instead the level of the civilian strategic reserve was established after the rest of the program had been decided upon.[q] It was viewed as a way to protect the imports that could be expected if the world price for crude oil were to fall to $7 a barrel in real terms.

The political need to include some specific conservation options, other than the price rises, in the program was met by relying mainly on subsidies and voluntary standards as the policy instruments. Voluntary standards were proposed for improving gas mileage, and for improving the efficiency of appliances. Mandatory thermal efficiency standards were proposed for new commercial buildings and homes. A bill to require auto and appliance manufacturers to place an energy efficiency label on all new automobiles and appliances was also proposed. Direct subsidies were suggested to assist low-income homeowners in insulating their homes and an income tax credit was proposed for certain stipulated household expenditures on improvements in residential thermal efficiency. Politically this part of the package was useful because it would not arouse the ire of any particular group, but it would demonstrate a willingness to go along with the politically popular notion that wasted energy must be reduced.

Thus, the Ford Energy Program was conceived. It was a compromise package designed to appeal to two audiences simultaneously. To the foreign audience it was presented as a tough program emphasizing the willingness of the United States to cut imports. To the domestic audience it was sold as a bold program to release the United States from the grip of OPEC. The process, at long last, was underway. The next move was up to Congress.

Notes

1. Theodore C. Sorenson, *Decision Making in the White House: The Olive Branch or the Arrows* (New York: Columbia University Press, 1963), pp. 25-26.

2. Cabinet Task Force on Oil Import Control, *The Oil Import Question: A Report of the Relationship of Oil Imports to National Security* (Washington: U.S. Government Printing Office, 1970).

3. William D. Nordhaus, "The Allocation of Energy Resources," *Brookings Papers on Economic Activity* 3 (1973): 532.

4. Timothy J. Healy, *Energy, Electric Power, and Man* (San Francisco: Boyd and Fraser Publishing, 1974), p. 140.

5. Council on Environmental Quality, *Environmental Quality: The Third Annual Report* (Washington: U.S. Government Printing Office, 1972), p. 243.

---

[q]The choice of a billion barrel strategic civilian reserve, as opposed to some other size, was apparently motivated by a preference for a round number. The proclivity of governments to fall back on round numbers in the absence of a scientific basis for any other choice is not restricted to energy policy. Note the parallel in designing the American Intercontinental Ballistic Missile in Morton Halperin, *Bureaucratic Politics and Foreign Policy* (Washington: Brookings Institution, 1974), p. 149.

6. Council on Environmental Quality, *OCS Oil and Gas: An Environmental Assessment* (Washington: U.S. Government Printing Office, 1974).

7. Dixie Lee Ray, *The Nation's Energy Future* (Washington: U.S. Government Printing Office, 1973), p. 15.

8. Richard Nixon, "The Energy Emergency: The President's Address to the Nation Outlining Steps to Deal With the Emergency. November 7, 1973," *Weekly Compilation of Presidential Documents* IX (July-December 1973): 1317.

9. *Congressional Quarterly Weekly Report*, 16 February 1974, p. 344.

10. Daniel J. Balz, "Energy Report/Summit Inflation Meetings Highlight More Questions Than Answers," *National Journal Reports*, 5 October 1974, p. 1503.

11. See Henry Kissinger, *The Troubled Partnership: A Reappraisal of the Atlantic Alliance* (Garden City, New Jersey: Doubleday and Co., 1966).

12. Henry Kissinger, "Interview with Newsweek Magazine," *Department of State Bulletin* LXXII (January-June 1975): 60.

13. Henry Kissinger, "The Year of Europe," *Department of State Bulletin* LXVIII (January-June 1973): 593.

14. Henry Kissinger, "An Age of Interdependence: Common Disaster or Community," *Department of State Bulletin* LXXI (July-December 1974): 503.

15. Henry Kissinger, "Major Oil-Consuming Countries Meet in Washington, D.C. to Discuss the Energy Problem: Statement by Secretary Kissinger," *Department of State Bulletin* LXX (January-June 1974): 203.

16. *Time*, 25 February 1974, p. 25.

17. Gerald Ford, "World Energy Conference: The President's Address to the Ninth Annual Conference in Detroit, Michigan, September 23, 1974," *Weekly Compilation of Presidential Documents* X (July-December 1974): 1181-86.

18. *The New York Times*, 5 October 1974, p. 3.

19. *Oil and Gas Journal*, 30 September 1974, p. 17.

20. *The New York Times*, 21 November 1974, p. 41.

21. Edward Cowan, "Ford's Energy Plan: Foreign Policy vs. Economics," *The New York Times*, 26 January 1975, section III, p. 2.

22. *Congressional Quarterly Weekly Report*, 12 October 1974, p. 2827.

23. Henry Kissinger, "Energy: The Necessity of Decision," *Department of State Bulletin* LXXII (January-June 1975): 237.

24. Richard S. Frank, "Energy Report/Ford Seeks Price Guarantees for Fuel to Aid U.S. Development," *National Journal Reports* 8 March 1975, p. 358-59.

25. Charles Bartlett, "A Slippery Floor," *Washington Star-News*, 27 February 1975; and *Baltimore Sun*, 5 February 1975.

26. *The New York Times*, 26 November 1974, p. 63.

27. Henry Kissinger, "Secretary Kissinger Holds News Conference at Brussels," *Department of State Bulletin* LXXII (January-June 1975): 1.

28. Daniel J. Balz and Joel Havemann, "State of the Union/Ford Pushes Program in Face of Strong Criticism," *National Journal Reports* VII (25 January 1975): 122.

29. Energy Independence Act of 1975, A Bill, S. 594, in the Senate of the United States, 94th Cong., 1st sess., February 5, 1975, pp. 97-100.

30. Finance Task Force, Federal Energy Administration, *Financing Project Independence, Financing Requirements of the Energy Industries, and Capital Needs and Policy Choices in the Energy Industries* (Washington: U.S. Government Printing Office, 1974), pp. II-15 to II-23.

31. Jerome E. Hass et al., *Financing the Energy Industry* (Cambridge: Ballinger, 1974), pp. 5-6.

8

Congress: Energy Initiatives and Responses

As an institution Congress is almost unique in its form and its powers among the legislative bodies of the world. The role that Congress plays in developing a policy framework is shaped by its constitutional authority, its rules and method of operation, and its political composition. In this chapter we examine the importance of each of these factors in shaping the congressional response to the need to establish a comprehensive and coherent energy policy.

The Constitutional Basis

Perhaps the easiest way to illustrate how the constitutional design of Congress shapes its role is to contrast it briefly with the British Parliament.[a] Under the British parliamentary system the chief executive is a member of Parliament, as are the other figures in his cabinet. All are elected by Parliament. He is prime minister solely by virtue of the fact that he is the leader of the political party that controls Parliament. His parliamentary party serves primarily to ratify and enact the decisions of the party leadership. Therefore, the distinction between the executive and legislative branches is very blurred, if not nonexistent.

Within this system party discipline and party loyalty are essential. Without this support the cabinet resigns and new parliamentary elections are held. Consequently party discipline within Parliament is extremely strict by American standards. Decision-making power is generally centralized in the leadership. This is not to say that it can impose its arbitrary will on the party. If it pursues policies consistently repugnant to the majority, the leadership may be replaced. The British party leadership makes a constant effort to stay in touch with party sentiment. However, as long as it stays within the general party consensus, the leadership controls formulation of policy and the legislative party simply enacts decisions.

This leadership-party relationship is reinforced by the fact that members of Parliament, while elected from a specific district, do not owe their principle

[a]The following section draws from Sidney D. Bailey, *British Parliamentary Democracy*, 2nd ed. (Boston: Houghton Mifflin Company, 1962); Samuel H. Beer et al., *Patterns of Government: The Major Political Systems of Europe*, 3rd ed. (New York: Random House, 1973), pp. 121-332; Lewis A. Froman, Jr., *The Congressional Process: Strategies, Rules and Procedures* (Boston: Little, Brown and Company, 1967); Louis W. Koenig, *Congress and the President* (Glenview, Illinois: Scott, Foresman and Company, 1965); and Peter Wall, *Public Policy* (Cambridge: Winthrop Publishers, 1974), pp. 162-202.

loyalties to that area. Members of Parliament may not be from their districts. Because they have been elected primarily to strengthen their party's position in Parliament, not to fight for regional interests, their constituency is a national one. In short, by American standards at least, the British parliamentary party is a national, cohesive, and dependable unit. While infighting is common, independent stands almost always give way to unity behind the national leadership's decisions when important issues arise.

The American Congress differs from its British counterpart on almost every one of the above points. The Constitution carefully separates the executive and legislative branches, giving to each unique powers. The President is elected in his own right, separately from Congress, to a fixed term of office. It is possible, and not uncommon, for the executive and legislative branches to be controlled by different parties. Congress, even when it is controlled by the President's party is not merely an enacting agent for the executive; it is a separate and co-equal branch of the government. The President must submit his proposals in the form of legislation to Congress. Only if it approves will they assume the force of law. Since Congress is not under the direct leadership of the President it may enact its own legislation, or substantially modify a presidential proposal. The President may veto a bill that has been passed by Congress, but this veto can in turn be overridden by a two-thirds majority of Congress. For legislation to pass (assuming that Congress can sometimes but not always override the President's veto) both branches must be willing to compromise with the other and arrive at a mutually agreeable solution. The effect, then, of the constitutional framework is to set up an adversary relationship between the executive and Congress, so that each may check attempts of the other to unilaterally enact policy.

This adversary relationship, when coupled with the difficulty of building a coalition of sufficient size to insure passage of a particular bill within Congress, makes the passage of new legislation difficult. This is, of course, especially true for controversial issues. Since Congress is the chief forum for the expression of regional interests, sharp differences in regional interests make the search for a compromise both more important and more difficult. If a deadlock results, no legislation may pass at all, in spite of the common desires of Congress, the President, and the public to take some sort of action. An example of how this can happen, even during times of intense concern over an issue, can be found in the fate of the Emergency Energy Bill of 1974.

President Nixon on November 7, 1973 had requested authorization to implement gas rationing and emergency conservation measures if needed. Senator Henry Jackson (D-Wa.) head of the powerful Senate Interior and Insular Affairs Committee, and a critic of the administration's policy, attached a number of amendments to the President's bill. These included a stringent windfall profits tax, congressional veto power over the President's conservation actions, benefits to those unemployed by the energy crisis, and loans to those engaged in energy-saving home improvements. These amendments were vigorously opposed by the administration, oil-state congressmen, environmentalist groups, and oil company lobbies. Under such pressure the development of a successful compromise bill proved difficult and the bill was twice recommitted to committee. Passage of a modified version was delayed until February 27, 1974. By that time Senator Jackson had abandoned the windfall profits tax, but had also rejected the administration's own milder version. He had instead incorporated into the bill a rollback in the price of all domestic crude oil to $5.25 a barrel. This was also considered unacceptable by the administration and its allies. Nixon vetoed the bill on March 6, and the Senate sustained his veto.[b]

All parties concerned wanted to take prompt steps to insure that the President would have adequate powers to deal with a severe shortage. Yet, severe differences in opinion not only delayed the bill but resulted in no bill at all. This example also is indicative of two other important factors.

First, it provides a glimpse into the prevailing ideology of Congress, which emphasized nonmarket over market allocation methods. During the period we are examining, Democrats generally did not want to achieve greater energy conservation by forcing the consumer to pay higher fuel costs, a direct contradiction of the prevailing administration view. Second, this bill illustrates the fact that Congress felt strongly that it should play a firm role in erecting an energy policy framework. Energy policy was potentially a platform that could be used to swing the balance of power between the legislative and executive branches back toward the legislative branch. We shall return to these themes in subsequent sections of this chapter.

The Congressional Role in Institutional Change

The second dimension in understanding how responsive an organization Congress is or can be has to do with its internal structure and how adaptive this structure to changing policy needs. Congressional approval in 1974 of administration bills creating the Federal Energy Administration (FEA), the Energy Research and Development Administration (ERDA), and the Nuclear Regulatory Commission (NRC) is indicative of congressional awareness of the need to restructure government bodies to deal with the new realities of energy, as well as of the desire to deemphasize nuclear research. Although Congress took no action regarding President Nixon's proposal to create a new Department of Energy and Natural Resources, the overwhelming margins by which the bills creating FEA,

[b]For accounts of the course of this bill, see Joel Havemann, "Energy Report/Emergency Legislation Dies; Opposition Gained, Support Waved," *National Journal Reports*, 5 January 1974, pp. 24-31; *Congressional Quarterly Weekly Report*, in the following issues: 2 February 1974, pp. 235-37, 252; 16 February 1974, p. 380; 23 February 1974, pp. 507-9; 2 March 1974, pp. 574-84; and 9 March 1974, pp. 632-33.

ERDA, and NRC were passed[c] serve as proof that Congress was attuned to the need for extensive changes in the structure of the executive branch.

However, the record of Congress in dealing with its own structural limitations, particularly as they related to energy, was much less impressive. Although Congress has been undergoing a slow, but profound, process of change and democratization, weakening the power of the committee chairmen, it has resisted all attempts to streamline the committee system in regards to energy legislation.

To examine the mechanisms of the congressional process in detail would require an entire book.[d] However, the most important elements for our present purposes can be quickly described. Once a bill is submitted to Congress it is channeled to the appropriate committee for consideration. In the House, for example, the Ways and Means Committee handles all tax legislation, the Foreign and Interstate Commerce Committee handles all legislation dealing with such trade, etc. The Rules Committee determines how a bill, which has passed through the appropriate legislative committee, shall be handled on the floor. It controls the amount of time to be given to debate, for example, or whether amendments may be attached. The same general process is followed in the Senate.

While the committee chairmen, elected by virtue of seniority, were not all-powerful prior to 1975, those heading the most important committees clearly possessed great power and, because of this, they were men who had to be deferred to. For example, the Rules Committee, under the chairmanship of Representative Howard Smith (D-Va.), hindered President John F. Kennedy's congressional program by simply refusing to send a number of his bills to the floor. The control of the committee by a bipartisan conservative group was sufficient to stop a number of liberal bills.[1] The Ways and Means Committee, until the recent reforms, was perhaps the single most powerful committee. Article 7 of the Constitution directs that all bills concerned with the raising of revenue must originate in the House of Representatives. Within the House the Ways and Means Committee has jurisdiction over tax legislation. The Ways and Means Committee was important for another reason as well. Because it controlled the appointments to committees, it was able to use these appointments as a bargaining lever to gain key votes on a particularly important piece of legislation. Other committees, although in general less powerful than the Rules Committee and the Ways and Means Committee, were nonetheless quite powerful in terms of bills within their jurisdiction. Most bills die in committee and those that emerge usually bear the strong stamp of the committee.

One problem associated with this powerful system of committees within

---

[c]The FEA bill passed the House 356 to 9 on April 9, 1974 and the Senate by voice vote on May 2. The ERDA-NRC bill passed the House 372 to 1 on October 9, and the Senate by voice vote on October 10. *Congressional Quarterly Weekly Report*, 19 October 1974, p. 2926.

[d]In fact, several such books have already been written. See, for example, Lewis A. Froman, Jr., *The Congressional Process: Strategies, Rules, and Procedures* (Boston: Little, Brown and Company, 1967).

Congress is that because multidimensional issues, such as energy, fall within the jurisdiction of several committees, the system breeds jurisdictional disputes. Prestige and visibility are usually associated with committees dealing with currently hot issues. Because these are very politically desirable attributes, members and chairmen are very sensitive about any infringements, apparent or real, on their legislative jurisdiction. For example, Senator Russell Long (D-La.), chairman of the Senate Finance Committee, and supporter of oil-price decontrol (Louisiana is an oil state), in July 1975 began to draw up a windfall-profits tax bill that would be acceptable to the oil industry. His reasoning was that such a tax would greatly increase the likelihood of congressional approval of a plan to decontrol oil prices. Representative Al Ullman (D-Ore.), Chairman of the Ways and Means Committee, who had previously favored such a combination of decontrol and a windfall profits tax, quickly made clear his opposition to such an intrusion on his jurisdiction. He indicated that if such a tax was necessary, his committee would be the one to draw one up.[2]

When neither committee in such a dispute has a clear right to a piece of legislation, and each is determined to establish or maintain its jurisdictional rights, the conflict can seriously delay action. A good example of this can be seen in the handling of Title I of President Ford's Omnibus Energy Bill, which permitted increasing the rate of production from the various naval petroleum reserves. This was an important component of most short-term energy strategies because it represented one of the few means of increasing domestic production in the period prior to 1977. As such, it offered a unique opportunity, albeit small, to reduce vulnerability in that crucial period without running the risk of increasing unemployment.

Originally, in late January 1975, Title I was assigned to the House Armed Services Committee. Then, in March, the House Interior and Insular Affairs Committee generated a bill that would transfer reserve control from the Navy to the Interior Department—overseen, of course, by the House Interior and Insular Affairs Committee. The Armed Services Committee protested by taking jurisdiction over the Interior Committee's bill and amending it so as to retain control. When the dispute finally moved to the floor, the House decided in favor of taking the jurisdiction away from Armed Services. The Interior Bill, with amendments restoring its original intent, was passed on July 9. Unfortunately, the situation was then complicated by a lack of coordination between the House and Senate. The Senate had, in the meantime, passed a version drawn up by the Senate Armed Services Committee on July 29, which maintained military control over the reserves.[e] This required a House-Senate conference, which, by the end of September, had not yet resolved the matter.[f]

[e]This account is based on Richard Corrigan, "Energy Report/House Committee Dispute Plays Action on Navy Oil Fields," *National Journal Reports*, 17 May 1975, p. 746; Elder Witt, "Development of Naval Reserves Approved," *Congressional Quarterly Weekly Report*, 12 July 1975, p. 1501; and Edward Dowan, "Senate Votes to Authorize Use of Naval Reserve Oil," *The New York Times*, 30 July 1975, p. 6.

[f]House-Senate relations on energy policy were even more difficult than usual during the first

This lack of coordination concerning energy bills in Congress is not confined to House-Senate matters. This problem can be found within each House, between committees. The committee jurisdictions were set up before energy became an important issue, politically or otherwise. Since energy policy involves a very broad range of areas and instruments, responsibility for such policy is severely fragmented in Congress. A total of 13 House committees and 10 Senate committees have some degree of jurisdiction over energy.[g] Several committees may be involved in a single energy bill, each one examining a particular facet. This is exactly what happened to President Ford's bill when he submitted it to Congress in late January. The Ford Energy Program, as transmitted to Congress, was, with the exception of the various tax proposals, embodied in a single comprehensive act, The Energy Independence Act of 1975. This centralization of the various parts of the program was deliberate, because it would serve to call public attention to the fact that each of the 13 interlocking titles was considered to be an integral part of an overall, coordinated plan.

Congress could not respond in kind. The bill was broken down into its individual titles and then sent, title by title, to four House committees and nine Senate committees. One title was sent to four different committees with overlapping jurisdictions.[h]

Because of this fragmentation and an apparent lack of any informal

half of 1975 because the Senate was preoccupied with resolving a disputed Senate election in New Hampshire. This issue was before the Senate fully seven months until July 30, 1975 when the seat was declared formally vacant and a new election was ordered. During this time there was an intensive filibuster to prevent Senate action by all 38 Senate Republicans, who were afraid that, if allowed, Senate Democrats would award the seat to the Democratic candidate. Six cloture votes to end debate, a Senate record, were unsuccessful as the Republicans were joined by 4 southern Democrats who were against cloture in principle. Energy policy, meanwhile was a secondary consideration. *Congressional Quarterly Weekly Report*, 2 August 1975, p. 1710.

[g]In the House, the committees include Government Operations, Interior and Insular Affairs, Interstate and Foreign Commerce, Merchant Marine and Fisheries, Public Works, Ways and Means, Armed Services, Banking and Currency, Education and Labor, Foreign Affairs, Judiciary, and Science and Astronautics. U.S., Congress, House, *Committee Reform Amendments of 1974: Report of the Select Committee on Committees of the House* (Washington: U.S. Government Printing Office, 1974), pp. 247-55.

In the Senate, the committees are Commerce, Finance, Foreign Relations, Interior and Insular Affairs, Labor and Public Welfare, Armed Services, Banking, and Housing and Urban Affairs. U.S., Congress, Senate, *Major Energy Related Legislation Pending or Acted on by the 93rd Congress* (Washington: U.S. Government Printing Office, 1974), pp. 3-71.

[h]Title VIII, dealing with the definition of energy facility needs and the establishment of procedures facilitating their development once approved was directed to the Senate Committee on Interior and Insular Affairs, the Senate Committee on Commerce, the Senate Committee on Public Works, and the House Committee on Interstate and Foreign Commerce.

The Ford bill went to all the Senate committees listed in footnote g, save Aeronautical and Space Sciences. In the House the bill went to Ways and Means, Commerce, Armed Services, and Banking. See Elder Witt and Tom Arrandale, "Energy Policy: 'Overestimating the Capability of Congress?'" *Congressional Quarterly Weekly Report*, 28 June 1975, p. 1343.

mechanism to circumvent the problem, it is very hard for an omnibus bill to proceed intact through Congress. Congress is not organized effectively to consider bills involving a variety of objectives and instruments. For example, Representative Ullman attempted to shepherd an alternative to the President's plan through his committee and the House during mid-1975. His original proposal included the group decontrol of oil prices, coupled with a windfall profits tax on oil companies. He had jurisdiction over a windfall-profits tax, but decontrol fell under the jurisdiction of the House Commerce Committee's Energy and Power Subcommittee, headed by Representative John Dingell (D-Mich.). Originally, Ullman and Dingell had intended to complete their bills simultaneously. However, the Dingell bill, which would have phased out controls over a five-year period, ran into intense opposition within the subcommittee from freshmen and liberal Democrats opposed to raising consumer prices. This opposition delayed the decontrol bill, and Ullman finally sent his energy bill to the floor without the accompanying windfall-profits tax.[3]

Observers and members of Congress are well aware of the problems inherent in the system. There have been countless proposals, and a few serious attempts, to reorganize Congress, or, at least, to centralize the congressional system for consideration of energy legislation. None have succeeded in permanently establishing a more coherent procedure.

In 1971 an attempt was made within the House to establish a single select committee to investigate energy resources. During the House debate a total of seven committee chairmen spoke up against the proposal on the grounds that such a committee would infringe upon the jurisdiction of existing units. The House concurred and defeated the proposal 128-218.[4] The size of the vote indicates the intransigence of the existing committee structure.

An attempt was made following the 1973 embargo to eliminate some of the delay involved in the committee system without attempting to reform the system itself. In June 1974 Congress completed action on a bill that included appropriations for all federal energy research and development programs. According to the traditional practice, such funding would have fallen piecemeal under the jurisdiction of six different subcommittees of the House Appropriations Committee, and so would have been broken up into six different bills. By incorporating all funding into a single bill to be scrutinized by the entire Appropriations Committee, the proposals were approved by the House and then by the Senate in a shorter period. This was only the second time such a procedure had been used in recent years. The immediate postembargo climate of crisis was apparently a prime factor in the fast action.[5]

Later in 1974, the House Select Committee on Committees, headed by Representative Richard Bolling (D-Mo.), recommended a drastic reorganization of the system. Many members of the Bolling group were especially concerned over the fragmented responsibility for energy legislation. Among other fundamental changes recommended, the committee report urged that jurisdiction over energy matters be centralized in a single House committee on energy and environment. The reorganization proposal was roundly denounced on the floor by important committee chairmen and discarded in favor of a much weaker alternative plan.[6] The rejection of the Bolling plan indicates that Congress could not institutionalize important changes in the committee system itself, in spite of the continued calls for change.[i]

However successful efforts to preserve the present committee system have been, a gradual but accelerating process of "democratization" has been at work in the House, steadily weakening the power of committee chairmen. The more autocratic committee chairmen, by the mid-1960s, had begun to draw the ire of their fellow congressmen. A series of reforms slowly loosened the hold the committee chairmen had exerted on legislation, and the congressional process became increasingly public, and increasingly accessible. In 1971 the power to vote on the retention of any given chairman was granted by the Democratic caucus to its members.[7] It was in 1975 that this power was first used to establish the power of the Democratic membership, in the form of the caucus, over the chairmen.

Since seniority was the criterion for selection, committee chairmen tended to be older and more conservative members, often from "safe" southern and rural districts. In November 1974 the more liberal character of the Democratic caucus was accentuated by the election of 75 freshmen Democrats. This more liberal caucus wasted no time; it promptly voted out three conservative chairmen. A fourth, Wilbur D. Mills (D-Ark.), chairman of the House Ways and Means Committee and for years the most influential chairman in the House, resigned his chairmanship at that time, due to personal problems.

The impact of the Democratic freshmen is discussed in greater detail in the following sections. Here it is sufficient to say that they accelerated the "democratization" of Congress. New reforms opened up more subcommittee positions to freshmen.[8] Mills was replaced by Representative Al Ullman, who is younger and less conservative. Ways and Means lost its power to make committee assignments, and was enlarged from 25 to 37 members including two freshmen. The increase in size made it harder for the chairman to control. The committee's deliberative sessions were now public, a move that did not please oil-interest lobbyists.[9] Most importantly, the caucus had demonstrated its power, and its willingness to use that power if chairmen attempted to circumvent the will of the membership.

For energy the first impact of this changing internal structure was felt when Congress, in a noteworthy reversal of its legislative history, repealed the oil depletion allowance. In 1974 Chairman Mills, who had historically been favorably

---

[i]For example, see Senator Mike Mansfield's comments, and the reactions of Representative Ullman, in Martha Angle, "Sen. Mansfield Proposes Streamlining of Congress' Energy Panel System," *Washington Star-News*, 22 May 1975.

disposed toward the oil-depletion allowance,[j] was nonetheless forced by congressional opinion to allow a tax bill including a gradual phaseout of the allowance to be produced by his committee. Under traditional House rules a Ways and Means bill could not be amended once it reached the floor. Two liberal members went to the caucus and got a vote directing the Rules Committee to allow them to attach amendments, one calling for an instant depletion repeal. Mills resisted this, and in spite of his already declining influence kept the bill from ever reaching the floor.[10]

The issue arose again in February and March of 1975 after Representative Ullman had assumed leadership of the committee. Committee Democrats opposing the allowance now strove to attach a repeal measure to the vitally important tax-cut bill, which they saw as being assured of passage. Ullman, although opposed to the allowance, refused to jeopardize the tax-cut bill's passage to any degree by attaching a controversial amendment. Again, committee members went to the caucus, which, now even more heavily inclined against the allowance, ordered the Rules Committee Democrats to ignore Ullman's request for a no-amendments rule and to permit the repeal amendment to be attached. Ullman was unable, and unwilling, to defy the caucus and backed down. The amendment was attached and the depletion allowance was repealed.[11] The repeal was a clear indication that the relatively liberal caucus was, if necessary, ready to outmaneuver and ignore committee chairmen in their attempts to dictate the form of legislation. In marked contrast to earlier days the House chairmen were no longer able to control the rank and file members of their committees, much less to control the fate of their bills on the floor.

All this is in accord with those preeminent principles of the American political system, pluralism and democracy. However, these reforms also had a second, more negative side. Power in the House had been further decentralized, and now that the chairmen had lost much of their power they could no longer insure and enforce a consensus. The nominal House leader, Speaker Carl Albert, had neither the power nor inclination to be such an agent. While the caucus itself was powerful, it was not under the firm control of any one group or individual. Furthermore, Democrats were not inclined to let any group, including the caucus, dictate their votes. The power vacuum within Congress made it difficult not only to form an internal consensus on controversial energy issues but it also hindered the process of forming a consensus with the President, since he no longer could talk to chairmen and believe that they spoke for their committees.[k]

[j]Mills had long been friendly to oil company interests. Along with many other legislators he had received contributions from major firms. In addition he had received an illegal $15,000 contribution from Gulf Oil during his abortive 1972 presidential campaign. See Congressional Quarterly Inc., *Continuing Energy Crisis in America* (Washington: Congressional Quarterly, 1975), p. 90.

[k]For elaborations on these themes see Donald Smith et al., "Overview: Democrats Worry

Viewed narrowly in terms of their effect on the responsiveness of Congress, the effect of the institutional reforms was, then, to leave intact the basic framework, with its drawbacks, while reducing its capability to force together the many interests represented in Congress into a consensus. This was to substantially weaken Congress in its attempt to initiate energy policy during 1975.

Congress as Policy Initiator

Congress has not always been the reacting rather than initiating branch. During America's history the initiative has swung back and forth between the executive and legislative branches.[1] After the depression, however, the balance of power drifted toward the President, although there have been some significant legislative initiatives in the intervening period. During the late 1960s and early 1970s a number of signs suggested that the time might be right for some reassertion of congressional power. The most important of these was the self-destruction of two presidents, Lyndon Johnson and Richard Nixon. Viet-Nam finally spurred Congress to reassert itself in foreign policy through the War Powers Act and by its cutoffs of aid to Cambodia, Viet-Nam, and Turkey. Richard Nixon had alienated Congress, and it succeeded in forcing him to resign when the Watergate scandal enveloped his administration. The fall of 1974 saw voters turn away from the Republican party and give the Democrats even larger majorities in both the House and Senate than before.[m] Among congressional Democrats, Gerald Ford was expected to be a relatively weak president and politically vulnerable. In sum, it appeared that Congress, emboldened by its recent triumph over Richard Nixon and its new Democratic majorities, might now have the strength and the will to reassert itself. Since energy was a highly visible issue, with voter concern quite high, it naturally became one of the areas in which the 94th Congress chose to make its stand.[n]

About 'Minority Rule,' " *Congressional Quarterly Weekly Review*, 28 June 1975, pp. 133-37, 1339-43; Elder Witt and Tom Arrandale, "Energy Policy: 'Overestimating the Capability of Congress?'," *Congressional Quarterly Weekly Report*, 28 June 1975, pp. 1343-46; James L. Sundquist, "Congress Braked by The System," *The Los Angeles Times*, 29 June 1975; and Bruce F. Freed, "House Democrats: Dispute Over Caucus Role," *Congressional Quarterly Weekly Report*, 3 May 1975, pp. 911-15.

[1]For a concise account of these swings see Warden Moxley, "Power Shifts Between President, Congress . . . A Basic Feature of U.S. Political System," *Congressional Quarterly Weekly Report*, 28 June 1975, pp. 1338-39.

[m]Following the 1974 election the Senate included 61 Democrats and 38 Republicans. One seat remained vacant because of the contested New Hampshire Senate election between Louis C. Wyman (R) and John Durkin (D). The House of Representatives was comprised of 289 Democrats and 144 Republicans. The death of two House members, Representative John C. Kluczynski (D) and Representative Jerry L. Pettis (R) resulted in vacancies in the 5th District of Illinois and the 37th District of California. Charles R. Brownson, *The Congressional Staff Directory* (Alexandria, Virginia: 1975).

[n]Congress had attempted to get a jump on energy policy in the enabling legislation creating the Federal Energy Administration, passed in May 1974. In that act the Federal Energy

Although the wide range of interests and ideologies represented by Democrats makes discussion of "Democratic goals" or "Democratic policies" difficult, it is clear that most Democrats objected to the adverse domestic economic consequences that could be expected to accompany the enactment of the President's program. In the first months of 1975 the nation's economy was sliding into the deepest economic downturn since the Depression. While unemployment soared, inflation continued to increase the cost of living. In the face of these problems with the 1973-74 oil embargo and price increases a year in the past, the American populace was not placing a high priority on resolving the energy crisis.[o] Gasoline was no longer in short supply and no future embargo appeared close at hand. As Representative Ullman was to put it, the energy crisis was an "invisible crisis."[12] The visible crisis was the state of the economy.

As a result of this feeling, the energy goals of most Democrats were at odds with those of the President. As a party they seemed little concerned with the role of domestic energy policy in shaping United States foreign policy, a main concern of the executive branch. Import vulnerability seemed to be a shared concern with the executive branch, but Congress gave it a somewhat lower priority.[p] None of the Democratic proposals, for example, involved cutting 1976 imports by 1 million barrels a day. All tried to mitigate, to some degree, the inflationary effects of dealing with energy while curbing unemployment. This represented more than the traditional Democratic identification with the interests of the poor, the workers, and the lower middle class. The special circumstances of the 1974 "Watergate" election were also responsible for increased Democratic sensitivity to voter concerns.

Many of the 75 Democratic freshmen had campaigned against big business interests, and against current government policies. Their election, as well as the increasing influence of consumer lobbies,[13] was an indication of an increasing emphasis on protecting consumer interests. Freshmen also had more than an ideological motive behind their advocacy of their constituents interests. Many freshmen had been elected from traditionally Republican districts.[14] If they wished to hold onto their seats in two years after the

Administration was required to come up with what was called in the act "The Comprehensive Energy Plan." The due date for this report was prior to the president's State of the Union Message, a traditional launching platform for new presidential initiatives. The preparation of this report was assigned a relatively low priority within FEA and was largely a collection of actions previously suggested by Presidents Nixon and Ford. There were numerous delays in preparing it and when it finally arrived in Congress it appeared to be treated with the apathy it deserved.

[o]FEA polls taken in December 1974, February 1975, and March 1975 asked sample groups to identify the number one problem facing them. Energy was selected by no more than 15 percent (February 1975) and by as little as 7 percent (December 1974). Unemployment was the public's major concern (from 50% to 57%) while inflation ran second (31% to 25%).

[p]In early February some important Senate and House members even questioned the need to cut oil imports at all. See Roberta Hornig, "Democrats Work on Plan Keeping Oil Imports Up," *Washington Star-News*, 2 February 1975.

Watergate backlash subsided, they would have to establish some credentials that would appeal to their constituencies. Consequently, the Ford energy bill was allowed to languish in the various committees, while Congress voted to suspend the President's authority to impose import fees on oil. Ford vetoed the suspension, but agreed to hold off on the imposition of the fees.

If stopping the enactment of the president's plan was not difficult, deciding who was to draw up the Democratic alternative was. With no group or individual in firm control a great number of plans and proposals arose.q Democrats were almost united in their opposition to the Ford plan, but they were far from united in deciding what alternative measures were required. This was reflected in their struggle to formulate a viable alternative.

The first Democratic energy proposal of the 94th Congress was the House Democratic leadership's plan, released on January 13.[15] A task force, headed by Representative James Wright (D-Tex.), had been named by Speaker Albert to create a Democratic economic proposal to counter the President's State of the Union Address. Because the task force was hopelessly hamstrung on the subject of energy by disagreements between factions, especially those involving environmentalists, however, the report was quite vague. It recommended mostly conservation actions for committee consideration. While it suggested an increase in gasoline taxes, the task force, sensitive to opposition to price increases within the Democratic party, treated the idea gingerly. Speaker Albert acknowledged the report's weaknesses and ordered the task force to try again.

It did so in conjunction with a Senate task force headed by Senator John Pastore (D-R.I.). The two forces developed their plans separately. Then, in a special conference, a joint plan was worked out. The primacy of economic stabilization over import vulnerability came through clearly in the design of the joint plan.

In this plan conservation was to be achieved primarily through a 5 cents a gallon increase in the federal gasoline excise tax, and a graduated tax on inefficient automobiles. To increase future supplies of energy the plan envisioned the creation of an Energy Research Trust Fund, which would presumably speed the development of environmentally sound alternative energy sources. To combat inflation and prevent excess profits in the oil industry an excess profits tax was suggested for large oil companies. "Old" oil prices would remain controlled at $5.25 a barrel. According to Representative James Wright, the entire package, if enacted, could be expected to reduce oil imports by 400 to 500 thousand barrels a day by 1976, less than half of the reduction sought by the President's program. If the measures failed to achieve desirable import levels the plan suggested the creation of a new National Energy Production Board, which could impose import quotas, not fees, and institute rationing.[16] The compromise plan, announced on February 27, 1975, was designed to have been

---

qBy mid-May over 1,250 energy-related bills had been introduced. *Business Week*, 26 May 1975, p. 26.

a program that the party could stand behind. However, it soon became clear that many Democrats objected to the plan, and that alternative plans would be presented.

On the one hand, more than half of the 75 Democratic freshmen did not like the compromise plan. They objected to the fact that the plan still involved an increase in prices, rather than a decrease, and they felt that the proposal would be ineffective. As an alternative to both President Ford's bill and the Wright-Pastore Plan, 40 freshmen in March introduced an energy plan based upon an increasing degree of government control over energy prices. It proposed mandatory enforcement of a national gasoline allocation system, limits on fuel price increases that could be passed along to customers by utilities, and a reduction in the prices charged by domestic oil producers.[17]

On the other hand, Representative Ullman objected to the Wright-Pastore plan because it did not raise prices enough. Ullman did not, by any means, want to allow prices to rise as much or as fast as the President did, but he did believe that the energy crisis was sufficiently important that a balance had to be found between protecting consumers from higher prices and curbing the wasteful use of energy. Early in March he presented his own energy proposal. It advocated a mix of market and nonmarket controls, differing from the other Democratic plans in two crucial areas.

First, the core of his plan was a gradually applied, but stiff, 40 to 50 cents increase in gasoline taxes. Second, he called for a phased-in decontrol of old oil and natural gas prices, in order to encourage production, coupled with a windfall-profits tax. Also included was a substantial excise tax on inefficient automobiles. Finally, to insure that imports be kept down, the draft suggested the use of quotas, if needed. This combined package would save, he estimated, 400,000 barrels a day in 1976.[18]

The Ullman plan had been formulated by members of the Ways and Means Committee. Various task forces made up of committee Democrats had submitted a series of recommendations. These recommendations were assembled into a plan by a new member of the committee who shared Ullman's views, freshman Democrat Representative Joseph L. Fisher (D-Va.). Fisher had little congressional experience, but had significant credentials in economics.[19] The resulting bill, while rather ingenious in its use of the price system, was not politically palatable to many segments of Congress. Nonetheless, the plan became the Democratic proposal, instead of the Wright-Pastore or freshmen plans. Although Ullman was not the powerful figure Wilbur Mills had once been, the Ways and Means Committee, by virtue of its jurisdictional prerogative over tax legislation, still exerted strong influence over energy policy.

In Congress the process of policy formulation is often difficult to separate from that of coalition building. The Ullman plan, presented in rough outline in March 1975 underwent a significant metamorphosis as Representative Ullman tried to obtain support for it both within Congress and with the White House.

This process of coalition formation influenced both the policy instruments to be used and the intensity with which they would be applied.

The Process of Coalition Formation

For a bill to become law, it must pass through a number of crucial decision points. A majority coalition favoring it usually must exist within the appropriate committee. Then one must be formed on the floor, and then the process must be repeated in the other House. Finally, unless Congress can be sure of its power to override a veto, presidential objections must be taken into account. These checks and balances tend to reinforce the status quo. A determined minority at any one of these crucial points can stop or modify legislation. The president can only stop, and not modify, with his veto, but the threat of a veto can induce modification beforehand. Opposition at each point can kill or radically weaken a controversial proposal.

It is during a bill's consideration by Congress that the variety of interests represented can express their views. Eventually, a compromise bill may emerge that is acceptable to a majority of congressmen. This process of compromise is desirable in that the final version should reflect divergent viewpoints. The difficulty with the energy issue for the 94th Congress was that nothing like a national consensus on what should be done about the energy crisis had emerged. Instead the members of Congress responded to intense constituent feelings as to what should not be done, that is, substantially raise prices and/or unemployment, or to hurt regional interests. Because Representative Ullman lacked the power to prod his fellow Democrats into a coalition, resistance from within his own party severely weakened his bill before it was finally passed on June 19, 1975.

Most objections centered on the gasoline tax increase. A group of over 100 of the House's 289 Democrats, including 4 committee chairmen and 50 freshmen, addressed a letter to Ullman in mid-March. In it they stated flatly, "We oppose increases in taxes on gasoline, whether sudden or gradual."[20] This intense opposition was reflected within the Ways and Means Committee as it considered its chairman's plan. As a result the scope of the tax was greatly reduced when the committee sent the bill to the floor on May 12.[r] The version sent to the floor called for an immediate 3 cents gasoline tax increase and an additional 20 cents increase to be phased in after the election year of 1976, if gasoline consumption were to increase beyond predetermined levels. The phase-in of

[r]The gradual weakening of the gasoline tax proposal can be seen in Peter Milius, "Ullman Unit Balks at 40 cents Gas Tax Rise," *Washington Post*, 12 March 1975; *Congressional Quarterly Weekly Report*, 22 March 1975, p. 580; 19 April 1975, p. 785; 26 April 1975, p. 859; Tom Arrandale, "Divided Panel Reports Energy Tax Bill," *Congressional Quarterly Weekly Report*, 17 May 1975, pp. 1016-17.

these taxes would have been automatic requiring no further discretionary action from either Congress or the executive branch.

Also weakened was the tax on inefficient vehicles. The fact that the American automobile industry was in serious trouble deterred the committee from imposing the severe taxes Representative Fisher wanted. Both the Ford Motor Company and the United Auto Workers lobbied forcefully against the Fisher version. It was replaced by a tax proposed by Representative Joe Waggoner (D-La.). The mileage standards set by the Waggoner tax were so close to what the automobile industry deemed feasible that no tax was likely to be assessed.[s]

Since these conservation clauses had been weakened, the sections authorizing import quotas grew in importance and became mandatory. However, the quota levels were raised by the committee, which did not want a shortage of oil to interfere with the economic recovery.[21]

The bill, thus weakened, was still unpopular among Democrats. It took all of Ullman's efforts to get it approved by the committee. The vote was close, 19-16, with four Democrats breaking ranks to vote with the Republicans to kill the bill. At the last minute, Ullman had to sacrifice a provision making the federal government the sole American purchasing agent of imported oil. Democrats from oil states objected to the clause and were apparently ready to kill the bill because of it.[22] The bill was also aided in winning committee approval by the feeling among some committee Democrats that they had to show something for their efforts. They doubted that the bill could be effective in its current state. However, not to pass it, after the repeated attacks on Congress by President Ford for failing to take action, would embarrass not just Representative Ullman but the committee as well.[23]

When the floor debate began it became clear that the Democrats, in spite of Ullman's compromises, would not approve his choice of instruments. The gasoline tax increase, seen as political suicide by many who had campaigned against such an increase,[t] was quickly dismantled. The 20 cents gasoline tax increase was overwhelmingly defeated 345 to 72.[24] Shortly afterwards the immediate 3 cents increase in the gasoline tax was defeated 209 to 187.[25] The tax on inefficient vehicles was also eliminated. The final committee version of the bill had contained a provision, the Waggoner Amendment, to tax the lower gas mileage cars of any manufacturer that failed to meet an average miles per gallon standard. The tax ran from 2 to 7 percent of the purchase price. When the bill hit the floor, in a last ditch attempt to strengthen the bill, Representative Fisher proposed an amendment to directly tax drivers, beginning in the 1977 year, who drove low-mileage automobiles. It was defeated 235 to 166.[26] Then

---

[s]The dilution of the auto efficiency tax plan, and the lobby pressures behind the process can be seen in *Congressional Quarterly Weekly Report*, 8 March 1975, p. 475; 3 March 1975, p. 953; and 10 March 1975, p. 960; and *Wall Street Journal*, 6 May 1975.

[t]As had Ullman himself. See Albert R. Hunt, "Two Achilles' Heels: Energy-Tax Bill Expected to Clear Panel, Then Stumble in Full House," *Wall Street Journal*, 12 May 1975.

the House accepted an amendment by Congressman Philip Sharp from Indiana, a state with great interest in the health of the auto industry, which would impose civil penalties rather than taxes on automobile manufacturers that did not prescribe the standards. This provision had been drafted by the House Commerce Committee and had been heavily supported by the automobile industry management and labor unions as an alternative to stiffer measures.[27] The amount of the proposed civil penalty was lower ($50 per car multiplied by the number of cars sold) than the tax it replaced but, more importantly, from the point of view of the industry, civil penalties were challengeable in the court (taxes are not) and could be lifted unilaterally by the President in case the industry could not meet the standards.

Three provisions of the bill survived, but the effectiveness of two of them was undermined. The import quotas were maintained, but only after they were further raised to levels above the expected import levels. The Energy Trust Fund, which had been designed to finance research in alternative energy sources, was retained, but it had lost its principal source of revenue, the 3 cents-a-gallon gasoline tax. A tax on the industrial consumption of petroleum, petroleum products and natural gas survived the floor fight intact.

The weakening of this bill resulted from a coalition of many diverse interests. For example, the automobile industry was effective in arguing that the various measures would mean unemployment in the automobile industry and because the auto industry was such a heavy buyer from other industries this would spread through the economy. Representatives from rural areas whose constituents rely heavily on the automobile for transportation because no mass transit alternative exists generally voted in large numbers against gas-tax increases.[28] Against these current and politically important arguments Representative Ullman had failed to establish a clear need in the eyes of the public.[u]

Regardless of the reasons, the bill as it left the House was less of an energy conservation bill than the House leaders had wanted and had expected. In spite of widespread initial enthusiasm for conservation among Democrats in principle (even the Pastore-Wright plan envisioned saving ½ million barrels a day through conservation by 1976), in practice the House found it impossible to find more stringent conservation proposals with which it could collectively agree. The House was severely criticized for its behavior by the press, by Republicans, and by the President. This criticism generally overlooked two important points: First, as indicated earlier, Congress cannot act in defiance of popular will. In this case it can be argued that the House performed exactly as was intended by the Constitution by protecting their various constituencies. Fear of defeat at the polls, a risk confronted every two years, was the mechanism used by the Founding Fathers to insure that legislation would reflect the will of the people.

[u]A Gallup poll taken just prior to the House consideration of the Ullman bill indicated that only 48 percent of those asked believed that an energy crisis existed while 44 percent believed that no such crisis existed. *Christian Science Monitor*, 12 June 1975.

It worked! Second, the bill was not completely ineffective, as argued in Chapter 9. In fact, because of the business tax provision if it were complimented by some additional legislation such as the implementation of the strategic reserve and by some measure to increase the domestic supply of oil it would, in some respects, be superior to the President's more ambitious proposal.

The process of coalition formation is rendered even more difficult by virtue of the fact that in order for the bill to become law the winning coalition either has to include the President or has to include two-thirds of both houses. Since the latter route is typically such a difficult one, throughout the committee hearings Representative Ullman and his supporters attempted to win and maintain White House support. The provision to include a tax on the business use of oil was in the bill for this purpose. It represented a move toward the President's position that conservation actions should not only be concentrated on gasoline, but should stimulate consumption reductions in all petroleum products.

Initially the White House gave the Ullman bill quite a bit of support, since it was the closest one to the President's own program, which was being actively considered in Congress. The FEA modeling capability was made available by the administration to the Ullman committee so that when various dilutions of the bill were proposed, all the members of the committee would be faced with the resulting rises in imports, which could be expected from any weakening amendment. But Ullman, unlike his predecessors, could not stray much further than he already had from the mainstream of Democratic feeling. Republicans in the end voted to kill the plan, and resisted attempts to ensnare their support for what they viewed as an ineffective bill that stood little chance of passage. The administration signalled House Republicans to withhold support by labeling the Ullman plan a "marshmallow."[29]

Although unsuccessful, Ullman's efforts indicate that he was well aware that any congressional energy plan would have to be at least marginally acceptable by the White House to become law. Unless Congress is certain of its override power this is true of any piece of legislation. Special circumstances in 1975 made the creation of a coalition between the two very difficult to obtain.

First, the Presidency and the Congress were not only controlled by men of different parties, but of different ideologies. Gerald Ford was a classic conservative, while the Democratically controlled Congress was especially sensitive to liberal and consumer interests. A complex issue like energy was bound to become heavily politicized. The President and Congress both felt the need to establish their positions before slowly attempting to compromise. This slow process was further complicated by the numerical deadlock that existed. Congress had the votes to stop President Ford's proposals. However, in spite of the President's talk during the 1974 election of a "veto-proof" Congress, it soon became clear that such an animal did not exist. In spite of their two-thirds majority, House Democrats found it almost impossible to override vetoes of congressional policies. With both branches determined to block the policies of the other, compromise became quite elusive. As a result, situations arose in which neither side was able to break the deadlock, and policies desired in substance by both the president and the Congress were not enacted.

The demise of the strip-mining bill provides a good example. A clear consensus existed that something should be done to curb the most environmentally destructive practices in getting the coal out of strip mines, although there was some disagreement on the degree of control to be applied and the method of implementation. Following a presidential veto of a strong strip mining bill in late 1974, Congress, in early 1975, perhaps confident of its supposed newfound ability to override vetoes, passed essentially the same bill. Again Ford vetoed the bill and Congress narrowly failed to override.[30]

Both sides attempted to break the deadlock by outmaneuvering the other. The Democrats attached the oil-depletion allowance repeal measure to a bill that the President was extremely unwilling to veto, the tax-cut bill of 1975. However, this time-honored technique of attaching a controversial rider to a veto-proof bill was not one that could be used frequently. There simply were not that many bills Ford wanted with sufficient urgency to make him accept undesirable amendments.

The President attempted to break the deadlock over the omnibus energy bill by indicating that he would explain the lack of enacted energy policy to the public as the sole result of a "do-nothing" Congress. He implemented this strategy by imposing a deadline on Congress, and calling public attention to that fact.

When Ford had agreed to hold off the imposition of additional import fees, following his veto of the suspension of his authority to do so, he gave Congress two months, or until May 1, to create an energy plan, or to pass his own. If it failed to do either, he threatened to impose the import fees. As a politically astute veteran of Congress the President must have guessed that the Democrats were too disorganized to respond. On May 1 he gave them another month. Again Congress failed to enact a plan. When the Ullman bill was introduced on the House floor in late May it had been immediately inundated with proposed amendments. Many of these were introduced by Republican members, possibly working to delay its consideration beyond the deadline. The House leadership surrendered and postponed further action until after Congress' return from a Memorial Day recess.[31]

In one sense, the president's tactic was a success. He publicly blamed Congress for the standstill on energy in a television address on May 27. Tearing off the months, page by page, from a calendar to emphasize the time Congress had "dawdled," he announced the imposition of the second set of import fees.[32] Yet, in a larger sense, President Ford failed. His address, and his successful vetoes of a number of congressional bills certainly served to increase the tension between the two branches. His deadline tactic had failed to generate a policy response from Congress.

Immediately following the passage of the modified Ullman bill on June 19, 1975 came the debate over the decontrol of "old" oil prices.[v] Ford had made decontrol a primary goal in January. Congress resisted an immediate increase, and Ford proposed a two-year phase out of controls. When the Dingel Energy and Power Subcommittee of the House Commerce Committee passed a five-year phase-out bill, it appeared that Ford and Congress would eventually strike a compromise. But the Dingel bill had passed through his subcommittee by a vote bitterly contested by Democratic freshmen, and it failed to survive in the full committee. The committee voted 22 to 21 to roll back old oil prices. An attempt was made to work out a compromise with Ford, but this attempt failed when the President refused to promise his signature for a compromise. Strengthened by this, committee Democrats toughened their bill and passed it 26 to 17. The proposal would have rolled back the price of new oil, and increased the price of old oil until they converged at a price between $7.50 and $8.50 a barrel.

The President refused to compromise then because he felt that he had sufficient power to force Congress to come around. He could veto any rollback bill, or even one simply extending the present controls for six months, and Congress would not be able to override him. Ford apparently felt that Congress would ultimately refuse to allow prices to jump on August 31, and so would eventually accept a compromise decontrol bill. Consequently Ford let it be known that he would veto an extension. In effect, he was again using a deadline to prod Congress into accepting his position.

On July 14, 1975 he outlined his compromise plan. The decontrol of old oil would be phased in over 30 months, and a ceiling of $13.50 was set on domestic price levels. Under the existing control law, either House or Congress could in effect veto the President's action within 5 days, and Congress was prepared to do so. The reasoning of many Democrats ran parallel to the President's. They believed that his unwillingness to accept public blame for a sudden price jump

---

[v]The following account is drawn from these sources: Gerald Ford, "The State of the Union: The President's Address Delivered Before a Joint Session of the Congress," January 15, 1975," *Weekly Compilation of Presidential Documents* XI (January 20, 1975): 49; Edward Cowan, "Congress, Ford and Price Controls on Oil," *The New York Times*, 6 May 1975; Edward Cowan, "House Panel Drafts A Bill To End Oil Price Control," *The New York Times*, 15 May 1975; Peter Milius "Commerce Unit Votes to Lift Price Controls as Domestic Oil," *Washington Post*, 15 May 1975; John Pierson, "Energy Measure Wins Approval of House Panel," *Wall Street Journal*, 25 June 1975; Rowland Evans and Robert Novak, "Ford's Energy Showdown," *Washington Post* 29 June 1975; *Wall Street Journal*, 11 July 1975; Carroll Kilpatrick and Peter Milius, "Ford Offers Oil Decontrol Compromise," *Washington Post*, 15 July 1975; Richard L. Lyons, "Hill Completes Oil Price Control Bill, Sends It To Ford for Probable Veto," *Washington Post*, 18 July 1975; Carroll Kilpatrick, "Ford Vetoes Oil Price Control Bill," *Washington Post*, 22 July 1975; David E. Rosenbaum, "Ford is Rebuffed By House Oil Vote," *The New York Times*, 23 July 1975; John Pierson, "House Rejects Twin Measures on Oil Prices," *Wall Street Journal*, 24 July 1975; Carroll Kilpatrick, "President Offers Hill 2nd Oil Decontrol Plan," *Washington Post*, 26 July 1975; *The New York Times*, 28 July 1975, p. 7; Edward Cowan, "Offer By Ford to Abolish 60 Cent Oil Fee Reported," *The New York Times*, 29 July 1975, p. 15; and Elder Witt and Tom Arrandale, "Oil Price Controls Extended: Veto Likely," *Congressional Quarterly Weekly Report* 2 August 1975, pp. 1655-57.

resulting from his veto action would finally lead him to at least sign an extension of controls. While continuing work on the Commerce Committee bill, Congress quickly passed a milder rollback bill that would have cut new oil prices from $13.00 to $11.28, extended the congressional veto period, and extended present controls for four months. On July 21 the President vetoed the congressional bill. On July 22 the House killed the President's plan. Having now publicly established their positions, the two branches presumably were in a position to compromise.

The House indicated its willingness to do so by voting down the bill rolling prices back to $7.50. But it also indicated the outside limits of its willingness by rejecting proposals for a gradual five-year phase out of old oil.

Ford then submitted his second compromise. Prices would be decontrolled over a 39-month period. A ceiling of $11.50 was set for all oil, although a 5 cents-a-month increase would be allowed. Ford urged that a windfall-profits tax be enacted to supplement his proposal. This new compromise, having been formulated after consultations with House Democrats, was carefully constructed to give them political freedom in approving it. Earlier, one House Democrat had said that voting for decontrol was like "applying for an exit visa from the Congress."[33] The new plan would decrease prices in 1975 and then increase them only slightly before the November 1976 election. The major increase would not have occurred until 1977. The White House gambled that a coalition of Republicans and conservative and oil-state Democrats would now be joined by moderate Democrats in sufficient numbers to pass the bill. As a final gesture, Ford offered to drop the 60 cents import fee placed on refined petroleum products if the plan was approved. If not, he again let it be known that he would veto an extension. Most observers and House leaders expected a narrow victory for the President.

Instead, the House voted, on July 30, to kill the President's plan, 228-189. The following day Congress sent a bill to extend the current system of controls to the president and the House renewed its plan to rollback new oil prices to $7.50. The new version of the rollback included one minor concession to the President. Oil coming from the Arctic Circle and from the Outer Continental Shelf, as well as oil recovered through the use of tertiary recovery techniques, could be sold at $10.00 a barrel. Following these actions Congress left on its August recess.

The President, living up to his threat, vetoed the extension of controls. On August 31, 1975 the legislative basis for the control system expired, but the oil companies, understandably uncertain about what would happen next, held back on the huge price increases they could no longer legally be denied. When Congress returned, a compromise was reached with the President to extend the current system of oil price controls until November 15, 1975, allowing one more chance to work a substantive compromise. Even this time proved insufficient as a further delay to December 15, 1975 was passed by Congress and signed by the President.[34]

Thus, the summer of 1975 passed without a resolution of the basic conflicts. Attempts at compromise had failed. The country did not have even the rudiments of a firm energy policy. As if in response to this policy vacuum the OPEC member nations approved a 10 percent increase in the price of their oil exports at their September meeting in Vienna.

In part this failure was due to the fact that Congress was now a more representative body than it had been in recent years, and, perhaps, ever had been. With such a diversity of opinion coalitions were hard to form. In part it was due to different underlying philosophies about what to do between the Republican-controlled executive branch and the Democrat-controlled legislative branch. In part it was due to the inability of the institutional structure to adapt in order to fulfill the organizational preconditions for managing the development of comprehensive and integrated energy policy.

The description of this process as it evolved raises a number of additional questions: How serious was this inaction? How desirable were the various programs, such as the Ford Energy Program, which were considered, but never translated into law? From the point of view of the energy policy objectives what policy choices would make sense? What kinds of hypotheses are suggested about the ability of the political system to deal with the energy problem in particular and other depletable resources in general? Part IV considers these questions and provides an assessment of the process and substance of energy policy.

Notes

1. Louis W. Koenig, *Congress and the President*, (Glenview, Illinois: Scott, Foresman and Company, 1965), pp. 47-48.

2. *The New York Times*, 1 August 1975, p. 1; and *Wall Street Journal*, 29 July 1975.

3. Peter Milius, "Ullman Panel Unveils Own Energy Plan," *Washington Post*, 3 March 1975; Edward Cowan, "House Panel Drafts a Bill to End Oil Price Control," *The New York Times*, 15 May 1975, p. C61; Tom Arrandale "Divided Panel Reports Energy Tax Bill," *Congressional Quarterly Weekly Reports*, 17 May 1975, p. 1016.

4. Congressional Quarterly, Inc., *Congressional Quarterly Almanac: 92nd Congress, 1st Session . . . 1971* XXVII, p. 799.

5. *Congressional Quarterly Weekly Report*, 4 May 1974, p. 1136; and *Congressional Quarterly Weekly Report*, 29 June 1974, p. 1707.

6. *Congressional Quarterly Weekly Reports*, 21 September 1974, p. 2578; *Congressional Quarterly Weekly Reports*, 28 December 1974, p. 3431; Congressional Quarterly, Inc., "Bolling Committee: Members Reforming House System" *CQ Guide to Current American Government* (Washington: Congressional Quarterly, 1974), pp. 68-71; and the Research and Policy Committee of the Committee for Economic Development, *Achieving Energy Independence* (New York: Committee for Economic Development, 1974), p. 66.

7. Donald Smith et al., "Overview: Democrats Worry About 'Minority Rule,' " *Congressional Quarterly Weekly Review*, 28 June 1975, pp. 1334-35.

8. Ibid., p. 1335.

9. Alan Ehrenhalf, "Energy Lobby: New Voices at Ways and Means," *Congressional Quarterly Weekly Report*, 3 May 1975, p. 942; and Bruce F. Freed, "House Democrats: Dispute Over Caucus Role," *Congressional Quarterly Weekly Report*, 3 May 1975, p. 915.

10. Daniel J. Balz, "Tax Report/Ways and Means Seeks to Maintain Power and Prestige," *National Journal Reports*, 22 June 1974, pp. 913-20; and IDEM, "Economic Report/Congressional Democrats Seek Program to Counter Ford," *National Journal Reports* 4 January 1975, p. 4.

11. Richard L. Lyons, "Oil Depletion Vote Nears," *Washington Post*, 26 February 1975; Eileen Shanahan, "Democrats Press for A Showdown on Oil Depletion," *The New York Times*, 26 February 1975; Richard L. Lyons, "House Unit Clears Way For Oil Depletion Vote," *Washington Post*, 27 February 1975; and John Fialka, "Liberal Clout on Oil Issue," *Washington Star-News*, 28 February 1975.

12. David E. Rosenbaum, "Energy Impasse: Lack of National Consensus," *The New York Times*, 22 May 1975, p. 25.

13. Ehrenhalf, "Energy Lobby," pp. 939-46.

14. Smith et al., "Overview," p. 1333.

15. Rowland Evans and Robert Novak, "The Achilles Heel of Congress," *Washington Post*, 13 January 1975; *Congressional Quarterly Weekly Report*, 18 January 1975, pp. 145-47; and *Congressional Quarterly Weekly Report*, 1 February 1975, p. 219.

16. Peter Milius, "Democrats Set Energy Plan," *Washington Post*, 28 February 1975; and *Congressional Quarterly Weekly Report*, 1 March 1975, p. 426.

17. Russell Warren Howe and Sarah Hays Trott, "The 'Watergate Babies'," *Saturday Review*, 31 May 1975, p. 11, 48.

18. Milius, "Ullman Panel Unveils Own Energy Plan;" Frances Anderson Gulick, "Energy-Related Legislation Highlights of the 93rd Congress and A Comparison of Three Energy Plans Before the 94th Congress," *Public Administration Review* (July/August 1975): 348-54; and Gene T. Kinney, "Demos Reveal Alternative Energy Plan," *Oil and Gas Journal*, 10 March 1975, pp. 26-28.

19. *Congressional Quarterly Weekly Report*, 8 March 1975, pp. 472-73; and Helen Dewar, "Freshman Fisher Has Key Role in Energy Proposals," *Washington Post*, March 5, 1975.

20. Helen Dewar, "Rep. Harris, Group Fight Gas Tax Rise," *Washington Post*, 16 March 1975.

21. See *Congressional Quarterly Weekly Report*, 12 April 1975, p. 752; and Tom Arrandale, "Divided Panel Reports Energy Tax Bill," *Congressional Quarterly Weekly Report*, 17 May 1975, p. 1017.

22. Arrandale, "Divided Panel," p. 1016.

23. Ibid.

24. U.S. Congress, House, 94th Cong., 1st sess., 11 June 1975, *Congressional Record*, pp. H5290-H5306.

25. Ibid., pp. H5290-H5325.

26. U.S., Congress, House, 94th Cong., 1st sess., 12 June 1975, *Congressional Record*, pp. H5350-H5376.

27. Helen Dewar, "Virginia's Freshmen and the Energy Tax," *Washington Post*, 13 June 1975; and Edward Cowan, " 'Gas' Tax Votes Reflect Mixed Needs," *The New York Times*, 13 June 1975.

28. David E. Rosenbaum, "House Rejects Bid for A Steep Rise in Gasoline Levy," *The New York Times*, 12 June 1975; Edward Cowan, "Congress Balks at Curving Energy Use," *The New York Times*, 12 June 1975; and Cowan " 'Gas' Tax Votes . . . ."

29. Rowland Evans and Robert Novak, "Playing Politics with Energy," *Washington Post*, 7 May 1975; and Roberta Hornig, "Ullman Accuses President of 'Bailout' on Energy Bill," *Washington Star-News*, 13 May 1975.

30. Carolyn Mathiasen, "Congress Sends Strip Mining Bill to Ford," *Congressional Quarterly Weekly Report*, 10 May 1975, pp. 964-67; Arthur J. Magida, "Environment Report/New Strip Mine Bill Ignores Ford's Suggestions," *National Journal Reports*, 8 March 1975, pp. 370-71.

31. Carroll Kilpatrick and Peter Milius, "Increases in Oil Tax Postponed by Ford," *Washington Post*, 5 March 1975, and Edward Cowan, "Ford Delays Oil Fee Rise, But Will End Price Curbs," *The New York Times*, 1 May 1975.

32. Peter Milius, "President to Add Second $1 A Barrel to Tariff on Oil," *Washington Post*, 28 May 1975.

33. Elder Witt, "Possible Break in Stalemate on Oil Prices," *Congressional Quarterly Weekly Report*, 26 July 1975, p. 1649.

34. *Wall Street Journal*, 13 November 1975, p. 16 and 17 November 1975, p. 8.

**Part IV:
Passing Judgment**

An Evaluation of
Alternative Policy
Scenarios

One senses in the United States a mood of frustration over the inability of our political institutions to act decisively in the face of the energy problem. It is inevitable, given the importance of the executive-legislative adversary relationship in explaining the lack of American decisiveness, that some will look wistfully toward the potentially more decisive British parliamentary system with its more harmonious executive-legislative roles and less fractured party loyalties. Since the desirability of decisiveness depends on the benefits conferred, it is important to assess the costs associated with indecisiveness and to reflect on the desirability of likely future outcomes from the American policy process as currently structured.

Although, as this book is being written, it is too early to state where the American policy process will ultimately lead, three rather important generalizations seem clear: First, the establishment of an energy policy promises to be a slow process. Delay is inevitable. Whatever energy policy emerges from this process is unlikely to affect the primary energy objectives in the next year or so. Second, Congress appears less willing to take stringent measures than the President and wants to phase the measures in somewhat more slowly. Finally, Congress seems to exhibit a marked preference for nonmarket policy instruments, while the administration insists on market mechanisms. The importance of each of these generalizations must be assessed.

This chapter provides a method for making that assessment. The analysis uses the estimates generated by the Project Independence modeling system in conjunction with some standard economic theory to determine the costs associated with several scenarios involving a range of possible outcomes from the policy process. The costs considered include the cost of a likely embargo in each year from 1975 to 1985 and the costs associated with the various government programs initiated to reduce this vulnerability. The final section of this chapter uses this information, in conjunction with other information, to suggest some possible resolutions of controversial matters, which might offer a reasonable series of compromises to break the executive-legislative deadlock.

Focusing this assessment on import vulnerability obviously leaves out some aspects of the energy problem and therefore should be defended. The effects of domestic energy policy on the balance of payments problem are excluded from the calculations, except to the extent that they overlap with import vulnerability, because they can be handled with complementary policies and are therefore separable. The balance of payments problem for the United States is expected to be of relatively short duration.[a] This country has the ability to substitute domestic oil resources for the expensive foreign ones and as a dominant exporter of food and industrialized goods it is expected to maintain an acceptable net export balance in spite of high oil import prices. The international short-term problem of insuring that the credit worthiness of the developing nations is maintained can be handled either by the International Monetary Fund or by special institutions set up to handle petrodollars recycling. Domestic energy policy, therefore, does not need to be addressed to this problem.

The foreign policy dimension of energy policy is decidedly trickier. It is neither separable nor quantifiable. The foreign policy benefits to be derived from particular policy choices are inherently more qualitative and amorphous. The increased solidarity of our alliance with Europe and Japan and the influence of domestic conservation policies on this solidarity is difficult to assess in anything but qualitative terms. Similarly, the effect of domestic energy policy on the ability of OPEC to maintain high prices is a difficult issue to handle with quantitative precision. Therefore, in this chapter the foreign policy dimension is discussed but no attempt is made to subject it to the same kind of quantitative assessment.

A Framework for Energy Policy Evaluation

A useful analogy in assessing the desirability of alternative policy packages in terms of their contributions to import vulnerability is the analogy of the individual buying insurance against an uncertain event, say having to rebuild his house after a fire. This individual will have to compare the costs of the fire insurance with the potential reductions in cost to him the insurance will bring if the fire breaks out. How much insurance the person should buy will depend on how likely it is a fire will break out, how costly repairing the damage would be, and the cost of buying the insurance.

The United States policy choices can be similarly viewed as an attempt to insure against future embargoes. Each policy package designed to provide this insurance, while reducing the potential costs of an embargo, also imposes costs on society. Strategic reserves, for example, have to be built and oil purchased. Higher energy taxes reduce real income by encouraging substitutions, which, in the absence of the tax, would be uneconomic. The nation must balance these two different types of costs.

Each policy package could result in a large number of outcomes depending on

[a]The net trade balance for the United States was positive for the six-month period ending June 1975 in spite of the high oil prices. See Council of Economic Advisers, *Economic Indicators* (July 1975): 1.

changes in significant factors beyond the control of the policy maker. Two sources of uncertainty stand out as being of prime importance. The first is the pricing strategy chosen by the Organization of Petroleum Exporting Countries (OPEC). As previous chapters have shown, this is an important determinant of the severity of the import vulnerability problem. If we know what strategy OPEC would pursue, United States policy choice would be somewhat easier. Since we do not, we are forced to hedge. The second source of uncertainty concerns the size, duration, and timing of any embargo that might (or might not) occur.

The evaluations in this chapter explicitly incorporate these elements of uncertainty. The exact treatment of them is given in Appendix A, but a flavor of the approach will be conveyed here. The general approach is to classify possible future outcomes and for each of these to characterize the likely costs associated with that outcome. These outcomes are then weighted by the likelihood of their occurrence and compared. In classifying these outcomes two OPEC pricing strategies are considered. The first involves a continuation of real $11 a barrel crude oil prices through 1985. The second assumes that the real price drops to $7 in 1978 and remains there through 1985. This latter scenario was the one used by the administration to plan and justify its energy program.

Several possible embargo situations are also considered. The costs associated with the different policy packages are compared for scenarios containing roughly one, two, or four embargoes. The embargoes are assumed to be equally likely in every year (i.e., the probability of an embargo in 1976 is assumed to be the same as the probability of an embargo in 1983), and each embargo is assumed to last six months, the approximate duration of the last embargo. The size of the embargo is determined by estimating the amount of imports that would come from embargo-prone sources in each given year for each given scenario and assuming that when an embargo occurs all these imports are embargoed.

The index of policy desirability used in this study is called the *present value of expected costs*; this represents the discounted stream of outcome costs weighed by their likelihood of occurrence. Since this number represents the expected costs the United States will likely incur, policy packages with lower values of this indicator are to be preferred to those with higher values because they represent a better balance of insurance and risk than the other packages.

The Costs of Inaction

As a point of departure, consider a hypothetical case in which the current political impasse was continued indefinitely except that the amount of old oil subject to price controls declined linearly to zero in 1985 and that natural gas was deregulated in 1975. In essence this is a base case in which market forces were unleashed by the government, but little else was accomplished. How serious a situation would this be? Estimates of the cost of a potential six-month embargo for each of several key years if this scenario would prevail are given in Table 9-1.

The explanation of why these costs behave the way they do is very straightforward. The costs associated with the most likely six-month embargo rise until 1977 because of increases in demand and declines in domestic production. The gap between demand and domestic supply increases over time and is filled by increases in imports. These imports are expected to come predominantly from embargo-prone nations. The numbers for both price scenarios for the first three years are the same by assumption since the real $7 price, if it occurs at all, is not assumed to occur until 1978.

The rising embargo costs in the 1975-77 period reflect not only increases in the size of the most likely embargo, they also reflect increases in the average cost per embargoed barrel. This results from a nonlinearity in the cost function, which implies that the cost of an embargo rises more than proportionately with the number of barrels affected by the embargo. Small reductions are achieved relatively easily and without great expense. Further reductions are more difficult and more costly.

The situation after 1977 changes because the Alaskan pipeline is expected to yield its first oil.[b] If the real world price were to drop to $7 a barrel, however, this reduction in import vulnerability would be of relatively short duration. More rapidly increasing demand coupled with more slowly increasing domestic supply, would eventually lead to higher imports and higher vulnerability.

The cost of an embargo in 1985, however, with continued $11 oil could be

Table 9-1

Estimated Costs of Potential Embargoes for Two World Crude Price Scenarios, Selected Years

(Billions of 1975 Dollars)

Cost of a Six-Month Embargo	1975	1977	1980	1985
$7 crude oil after 1977	$17.5	$38.3	$49.5	$154.7
$11 crude oil after 1977	17.5	38.3	14.0	0.0

Source: Calculations by author based on methodology and data described in the appendixes to this book.

Note: In this table and those that follow in this chapter both price scenarios assume $11 oil through 1977. The oil prices are real 1975 prices. The nominal prices would be expected to rise with inflation.

---

[b]This illustrates the importance of the decision to go ahead with the Alaskan pipeline, which finally cleared Congress during the 1973 embargo after a long period of indecision. This oil represents the only major contribution to reducing import vulnerability that will likely occur in the period through 1978.

expected to be quite small. Assuming, for a moment, the correctness of these numbers, a couple of important conclusions follow: (1) the costs of potential embargoes in the period from 1975 to 1977 are quite high, and (2) in the longer run continued high prices (i.e., maintaining the current real price), by restraining demand and stimulating domestic production, could reduce vulnerability to acceptable levels with no additional government action whatsoever. The important implication of this is that as long as OPEC maintains real crude oil prices at or above the $11 level and American industry expects this outcome and reacts accordingly the long-run cost of a political impasse between the President and Congress would not be very large. In other words, in this case the high prices initiated by OPEC could be expected to unleash the market forces to such an extent that the problem of political nonresponsiveness of institutions would not be particularly severe. If the world price were to fall to $7 or lower, however, by 1985 import vulnerability could be a severe problem without some form of insurance.

One should be suspicious of dogmatic conclusions about the future because so many uncertainties are involved in forecasting the outcomes on which the conclusions are based. The conclusions in this chapter are no exception. We can, however, illuminate some of the elements of this uncertainty and examine the cost of being wrong.

The demand elasticities used in the Federal Energy Administration (FEA) energy model, which generate the inputs for the analysis in this book, are quite price responsive. The degree of confidence in these numbers, however, is not large because current prices clearly are well out of the range of the data used to estimate the elasticities. If supply and demand tend to be less responsive to price than projected here, both the $7 and $11 outcomes will be characterized by higher embargo costs than used here. Conversely, if they are more responsive the corresponding embargo costs would be lower.

How should policy makers react in the fact of this uncertainty? In principle, since more information will be available on these uncertainties in the future, the government should preserve some flexibility in making future choices by using hedging strategies, with reevaluations of the situation designed to take place every few years. This would allow time to verify the price responsiveness of demand and supply and discover the price strategies of OPEC. In practice, this may be difficult because the policy process is so cumbersome and unresponsive.

A prime hedging strategy, the one favored by most economists, would be accumulation of an oil stockpile or strategic reserve in salt domes, which could replace the embargoed oil on short notice. By decoupling the relationship between the size of an embargo and the cost it imposes on the nation, stockpiles can lower vulnerability while still permitting the importation of foreign oil. The ability to use the foreign oil would, of course, be particularly valuable if the world oil price drops. Stockpiles lower the rate of depletion of scarce domestic oil reserves by diminishing the need for reducing the dependence on foreign supplies.

The potential impact of stockpiles on import vulnerability can be illustrated by hypothetically assuming that the 1 billion-barrel stockpile, as proposed by the President, were authorized in 1975. The estimated costs of embargoes from taking this action and only this action are given in Table 9-2.

How effective any storage program could be would depend crucially on its drawdown rate during an embargo. It is unlikely that the strategic reserve would be drained during any embargo for the simple reason that the ultimate duration of that embargo would be unknown until it were over. In addition, since the marginal cost to the economy rises with the number of barrels embargoed, it makes economic sense, even if the duration were known, to spread the drawdown over time. This economic rationale for spreading the use of the stockpile over time, coupled with the uncertain end of the embargo, leads to the conclusion that it is unlikely that the stockpile would ever be fully used during any single embargo. To account for this fact the estimates in Table 9-2 reflect a drawdown rate, which exhausts only 75 percent of the storage by the end of the assumed six-month embargo. This appears to be a reasonably realistic assumption.

The main conclusion conveyed by Table 9-2 is that, in the short run, storage does not make any contribution to reducing vulnerability at all, but in the long run it can make a substantial contribution. The limitations of storage in the short run have to do with the lead times necessary to leach the salt domes, to acquire the oil and pump it in, and to establish the logistical arrangements required to establish a high flow rate out of the reserve in case of an embargo. These steps take time.

Table 9-2
Estimated Costs of Potential Embargoes for Two World Crude Price Scenarios with a One Billion Barrel Storage Program

(Billions of 1975 Dollars)

Cost of a Six-Month Embargo	1975	1977	1980	1985
$7 crude after 1977				
Base case	$17.5	$38.3	$49.5	$154.7
With storage	17.5	38.3	11.6	11.4
$11 crude after 1977				
Base case	17.5	38.3	14.0	0.0
With storage	17.5	38.3	0.0	0.0

Source: Calculated by the author using the data and methodology in the appendixes.

Note: Neither these estimates nor the estimates in the tables that follow include the effects of participation in the International Energy Plan Sharing Agreement, since the specifics of this agreement are not clear at the time of this writing.

The inability of storage to affect vulnerability in the short run, however, should not obscure its tremendous contribution over the longer run. If the OPEC countries were to maintain a continued $11 real price for crude oil, and if the estimated demand and supply elasticities are essentially correct, then the United States would be essentially invulnerable to an OPEC embargo by the early 1980s without ever engaging in such controversial policies as the accelerated development of the Outer Continental Shelf. Even if the price were to fall storage would reduce vulnerability to below current levels by about 1980.

There are costs associated with the establishment of a strategic reserve and the decision whether to build such a reserve would have to consider these along with the benefits. The oil has to be purchased and the supporting facilities constructed. There is also some chance that the stockpile would never be used. Whether the benefits of a stockpile appear to outweigh the costs can be determined by the comparison of the present values of expected costs both with and without a stockpile. Assuming a 50-50 chance of continued high prices and two embargoes during the next 11 years leads to a present value of expected costs of $51.8 billion without the stockpile and $26.0 billion with the stockpile. It appears to be a very desirable hedging strategy. Unfortunately, as of November 1975 the law authorizing this stockpile had not yet been enacted. Its desirability seems so obvious, however, that passage is quite likely. Therefore, the remaining programs are compared to a base case with a 1 billion barrel storage program.

Comparison of Executive and Legislative Programs

A program based solely on storage is a fairly low profile program. It makes no attempt to seek self-sufficiency. It avoids many of the large controversies surrounding many of the policies, such as the accelerated development of the Outer Continental Shelf, the accelerated development of nuclear power, and the imposition of tariffs. The question then is, "What are the benefits to be accrued from taking these more powerful steps?"

To provide this kind of information two versions of two particular policy outcomes are considered. The mildest one in terms of its impact on import levels is represented by adding the policies in HR6860, the Ways and Means Committee Bill as passed by the House of Representatives in June 1975, to the base case including a strategic reserve. The third policy package, the President's program, represents the maximum assault on imports because it adds a tariff and a program of accelerated development of nuclear energy and of the oil reserves on the Outer Continental Shelf.

To round out the comparison two modifications of these policy packages are also included. The first represents augmenting HR6860 with a program of accelerating the development of domestic resources. The second is the President's program without the tariff. The latter is included primarily to show how costly a decision including the tariff in this package was.

This comparison serves two purposes: (1) it illustrates the costs and benefits to be derived from following more aggressive policies, and (2) it illustrates what could have been the outlook if either the President or Congress had been able to impose unilaterally its will without fear of contradiction by the other branch. At the risk of considerable oversimplification it provides a hint of what might have occurred if United States policy were formulated within a British type parliamentary system.

The first step in our evaluation is to portray the consequences of each of these two scenarios (and the base case for comparison) on import vulnerability, ignoring, for the moment, the costs associated with implementing those policies. The results of using the methodology and data in the appendixes to compute the cost of a representative six-month embargo for each of several selected years is presented in Table 9-3. To provide a standard of comparison the model being used estimates that a comparable type of embargo in 1973 to the ones considered here (i.e., embargoes in which the prices fall back to preembargo levels after the embargo is lifted) would have cost around $12 billion.

Table 9-3 makes very clear the contribution to import vulnerability that can be made by more ambitious policies. They can reduce the near-term vulnerabil-

Table 9-3
Estimated Cost of a Six-Month Embargo for Each of Four Policy Packages and a Base Case for Two World Crude Price Scenarios, Selected Years
(Billions of 1975 Dollars)

	$7 Crude Oil After 1977				$11 Crude Oil After 1977			
	1975	1977	1980	1985	1975	1977	1980	1985
Base case with storage	$17.5	$38.3	$11.6	$11.4	$17.5	$38.3	$0.0	$0.0
HR6860 with storage	17.3	34.6	7.8	4.3	17.3	34.6	0.0	0.0
HR6860 with storage and accelerated development	17.0	31.0	1.6	0.0	17.0	31.0	0.0	0.0
President's program without tariff	16.0	20.0	3.7	0.0	16.0	20.0	0.0	0.0
President's program with tariff	16.2	13.9	0.0	0.0	16.2	13.9	0.0	0.0

Source: Calculations by author based on methodology and data given in appendixes.

ity until the additional oil from the Alaskan pipeline becomes available and they can reduce longer term vulnerability if the efforts to get OPEC to reduce the real price of oil are successful. The table also makes clear, however, that if OPEC is able to maintain continued high prices, in the long run, these more ambitious policies make very little contribution to import vulnerability. This conclusion stems from the demand restraining and supply expanding characteristics of continued high prices in addition to the availability by 1980 of a substantial portion of the 1 billion barrel strategic reserve. The differences between the least aggressive and the most aggressive policy packages is marked. Is the pursuit of a more high profile energy policy worth it? The answer, of course, depends not only on the benefits to be derived from the package (i.e., reduced vulnerability), but also the costs incurred in implementing the policies.

There are several different costs involved. The first, shared by all strategies considered above, is the cost of the strategic reserve. The second is the cost of imposing higher prices on consumers than would otherwise prevail. Two costs, environmental costs and the costs associated with increasing the rate of depletion of domestic resources, are omitted because no reasonable estimates of them exist. The quantification of the aggregate of the two former costs is given in Table 9-4.

Table 9-4
Estimated Annual Costs of Alternative Policy Packages for Selected OPEC Pricing Strategies, Selected Years

(Billions of 1975 Dollars)

Strategy	$7 Crude				$11 Crude			
	1975	1977	1980	1985	1975	1977	1980	1985
Base case with storage	$0.10	$0.30	$1.40	$0.01	$0.10	$0.30	$2.10	$0.01
HR6860 with storage	0.11	0.33	1.46	0.25	0.11	0.33	2.20	0.25
HR6860 with storage and accelerated development	0.11	0.33	1.46	0.25	0.11	0.33	2.20	0.25
President's program without tariff	0.26	0.68	1.50	0.01	0.26	0.68	2.31	0.01
President's program with tariff	0.54	1.54	2.49	0.92	0.54	1.54	3.17	0.52

Source: These costs are derived by summing the annual costs of the strategic reserve and the cost to consumers of government-initiated price increases. The method for calculating each component is described in appendix A.

These figures point out the rather large costs associated with the President's program as it was submitted to Congress. Why was the program so costly? Continued controls on "old" oil would yield a current average price of crude oil of about $8.44 a barrel. This is due to the fact that the average cost represents a blend of the controlled domestic crude, the uncontrolled domestic crude, and the foreign crude. With immediate decontrol and a tariff this oil would all cost the same, approximately $13.00 a barrel. Thus, the $2.00 tariff actually leads to a $4.56 increase in the average cost of oil when coupled with the decontrol of controlled oil prices. This represents slightly over a 50 percent increase in the average cost of crude oil. Given the importance of oil in the American economy that is not an insignificant increase for the economy to absorb all at once.

To see whether these additional measures reduce costs, the present values of expected costs must be compared. Our comparison will explicitly consider three possible likelihoods of OPEC pricing strategies and three possible likelihoods of embargoes for each policy package. The interaction of these factors means that there will be nine calculations accomplished for each package. The uncertainty imposed on the United States by an inability to forecast with precision the OPEC pricing strategy will be treated by examining each package under the assumptions that the probability of OPEC choosing to drop the price to $7 in 1978 is respectively 0.25, 0.50, and 0.75. The uncertainty about the timing of the embargo is treated by assuming that the probability of an embargo is equal for all 11 years. Probabilities of 0.10, 0.20, and 0.40 are assumed. This roughly corresponds to an assumption that we have 1, 2, or 4 embargoes during the next 11 years.[c] Both of these assumptions would appear to cover a reasonable range of outcomes. The results are presented in Table 9-5.

There are several interesting facts to be gleaned from Table 9-5.

1. The widely criticized House bill, when supplemented by a strategic reserve and a program to accelerate the development of domestic production, is actually superior to the President's program if we have only one embargo in the next 11 years.
2. The decision to add the tariff to the President's program would have been a costly one for the nation if the program had been fully implemented. It makes economic sense only if the country is subjected to four or more
. embargoes in the next 11 years.
3. The various policy packages make the present value of expected cost rather insensitive to OPEC pricing policies. They are, in fact, hedging strategies.

These observations provide a basis for discussing both the substance and the process of energy policy.

[c]The exact percentages would be $1/11 = 0.09$, $2/11 = 0.18$, and $4/11 = 0.36$. The bias introduced by interpreting the above percentages in integer embargo terms is therefore very small. The direction of the bias is to make the more aggressive policies appear somewhat more favorable than they actually are.

Table 9-5

The Present Values of Expected Costs for Alternative Policy Packages, Selected Years

(Billions of 1975 Dollars)

Policy Package	Probability of a Disruption	Probability of a Sustained Fall in Price in 1978		
		0.25	0.50	0.75
Base case with storage	$0.10	$15.7	$16.1	$16.5
HR6860 with storage	.10	15.8	15.7	15.5
HR6860 with storage and accelerated development	.10	14.2	13.8	13.4
President's program without tariff	.10	13.2	12.9	12.7
President's program with tariff	.10	17.2	16.8	16.4
Base case with storage	.20	24.7	26.0	27.3
HR6860 with storage	.20	24.3	24.6	24.9
HR6860 with storage and accelerated development	.20	21.1	20.8	20.6
President's program without tariff	.20	18.3	18.4	18.4
President's program with tariff	.20	20.9	20.6	20.3
Base case with storage	.40	42.5	45.7	48.9
HR6860 with storage	.40	41.2	42.3	43.5
HR6860 with storage and accelerated development	.40	34.9	34.9	35.0
President's program without tariff	.40	28.3	29.1	30.0
President's program with tariff	0.40	28.2	28.1	28.1

Source: Computed from the data given in Tables 9-3 and 9-4 as well as unreported data for the intervening years using the methodology discussed in appendix A.

Note: The three programs, which include increasing the rate of depletion of domestic reserves (the modified HR6860 and the two Presidential programs), do not include any estimate of the increased associated environmental costs or the increased cost due to the fact that this depletion will hasten future scarcity. Therefore these costs are biased downward to an unknown extent.

The political give and take within the Congress and between the President and Congress as this book is written has failed to yield an energy policy. As was argued in earlier sections this could be quite costly. However, by the enactment of a limited number of policies the United States can set in motion forces that will provide a good deal of protection and will buy time while we gain more information about the position of OPEC. We need not make irreversible commitments now because the cost of delaying until we have more information is not very large, providing the strategic reserve is in place.

The Ways and Means Committee energy conservation bill was roundly criticized by the leadership of the House, by the administration, and by the press. In comparison to the version that was passed by the Ways and Means Committee, it was definitely a weaker and less desirable bill. Yet, it is incorrect to carry this charge too far and allege that the bill would have no impact. The tax provisions of the bill directed toward reducing energy consumption in industry were estimated by the FEA energy model to have a not insignificant impact on energy consumption. As Table 9-5 indicates the additional costs to industry imposed by the tax would be matched by a more than compensating fall in import vulnerability. Standing alone it is clearly not enough. Yet, when coupled with a strategic reserve and an accelerated development of domestic resources, both of which are elements of the President's program, it yields a present value of expected costs that is lower than the complete President's program if only one embargo or less occurs in the next 11 years.

These figures indicate that the President's program, as proposed, when examined by the strictly economic criterion of whether it lowers the expected present value of future costs, went too far. The move toward self-sufficiency, which it advocated, was unnecessarily ambitious. Yet, as was shown above, the program was designed to do more than simply minimize the costs of an unknown and uncontrollable OPEC-pricing strategy. It was designed to affect the OPEC strategy in two ways: (1) by providing the basis for increased consumer nation solidarity as a precondition for negotiating the price downward, and (2) by reducing demand sufficiently to put great stress on the ability of OPEC to allocate among its members the cuts in production that would be necessary to keep the price up. What is the evidence that the program could be expected to accomplish these objectives?

The available evidence does not provide much support for the Kissinger approach. Consider briefly each of these points. The hope for a powerful group of oil consuming nations to form a strong negotiation base does not appear to be realistic. The consuming nations, for example, have not been able to agree on a common approach to the price floor issue. In a recent paper David B. Bobrow and Robert T. Kudrle argue that this is to be expected[1] since the preconditions for a successful coalition are not to be found in the group of consuming nations. They are too heterogeneous in their dependence on Middle Eastern oil and in their role in the Arab-Israeli conflict. The United States, as a major producer of oil and as a major force in the Arab-Israeli conflict, does not have the same interests in negotiating a settlement as, say, Japan. The heavily dependent nations have repeatedly opposed joining any confrontation group.

Bobrow and Kudrle also point out that because of the internal politics within the two groups of countries it is more likely that oil exporters, singly and through OPEC, would manifest more coherent and flexible policies than the importing nations.[2] The preceding chapters have provided a case in point. There is not exactly a uniformity of opinion within this country on what should be done, much less within a diverse group such as the consuming nations. The executive-legislative differences and the Kissinger-Simon differences have made it difficult for this country to present a unified front. This absence of a unified domestic position makes the job of convincing other nations to go along that much more difficult.

Could it be, however, that a unilateral policy pursued by the United States, which amounted to an aggressive withdrawal of the United States from the world petroleum market, would put sufficient pressure on OPEC to lead to a price fall? The answer, both in the short run and the long run appears to be negative. In the short run because of the very low price elasticity of demand, very large price increases are necessary to reduce demand significantly. The best that could be hoped for was an additional 1 or 2 million barrels a day in a couple of years from what would have otherwise been demanded. But this would certainly be partially counteracted by the economic recovery as rising incomes lead to increased demand. The estimate by the Federal Energy Administration was that the predicted 1977 import level could be expected to be only about 1.1 million barrels lower than the 1973 import level.[d] This would not, by itself, make much dent in OPEC production, which was estimated at 34.8 million barrels a day in 1973.[3]

In the long run demand is much more elastic, but even in this case the influence of actions by the United States on OPEC production is likely to be small. If oil were to remain at $11 a barrel in real terms that would, by itself, call forth a drastic reduction in imports by the consuming nations from OPEC from what would have been imported at lower prices. Under the President's program with continued $11 òil the United States would be self-sufficient in petroleum by 1985. Yet, by doing absolutely nothing, it can be expected to import only about 3 million barrels a day. Will a 3 million-barrels-a-day reduction be sufficient to get OPEC to lower its price? With an expected production capacity of 53 million barrels a day,[4] it is not likely. The inevitable conclusion is that as large a force as the United States is on the world petroleum market its choice of strategy is not likely to affect OPEC's strategy. The price of oil may well fall, but whether this occurs will be largely independent of the policy chosen by the United States government.

Evaluating the Policy Issues

The foregoing analysis is quite suggestive with respect to the possibilities for future compromise. The fact that the President's program was more stringent than

---

[d]This information is in a memo from William W. Hogan, director of the Office of Quantitative Methods in the Federal Energy Administration to Jim Wetzler, staff economist for the Joint Committee on Internal Revenue Taxation, dated April 7, 1975.

necessary to accomplish the basic energy objectives suggests that he should be willing to give ground on this issue. On the other hand with all the uncertainty about what the future holds and the need for flexibility Congress should be somewhat more willing to unleash the price system than they have been. In this way political responsiveness and the responsiveness of the economic system could go hand in hand. Whether this will occur remains to be seen, but there are clear grounds for compromise. With this general thought in mind we turn to an evaluation of particular politically charged issues to see how they might fit into this general pattern.

The highest policy priority should be starting immediately on the development of a strategic reserve. With a reserve in place we could lessen considerably the pressure that OPEC can place on the United States in its foreign policy and we could buy time. This time would allow the accumulation of more information on how big the Outer Continental Shelf reserves are, how sensitive demand and supply is to prices, and how stable the OPEC cartel is likely to be. The increased time would also permit a more careful consideration of the more imponderable decisions such as the role to be played by nuclear power.

The second major decision concerns what to do about domestic oil prices. There are several dimensions to this problem. Currently there are by definition two kinds of domestically produced oil, "old" oil and "new" oil. Old oil has a controlled ceiling price of $5.25 a barrel. The determination of which oil is old oil depends on the property the oil comes from and the level of production from that property. For a particular property, which is not a stripper-well lease,[e] the volume of controlled oil equals the total number of barrels of domestic crude petroleum produced from that property in the corresponding month of 1972 minus an amount of released oil equal to the new oil production from that property.[f] All other oil is allowed to seek its own price.

There are three basic decisions to be made: (1) the appropriate price for new oil, (2) the appropriate price for old oil, and (3) the definition of old oil. The price for new oil is a driving force behind new discoveries of oil. It will also be a major determinant of the speed in which oil substitutes, such as solar energy, would be adopted. For this reason it should approximate the landed imported crude price, exclusive of tariff, because that is the oil it will replace.

The old oil dimension of the problem has two components. First, given the current system the incentive to introduce new recovery techniques is limited. Second, the current regulations actually encourage the importation of oil, which, of course, was not the intent of implementing them. Consider each of these arguments in turn.

There is very little question that truly old oil can be produced profitably at

[e]A *stripper-well lease* is a property that produced an average of less than 10 barrels of oil a day for the preceding calendar month.

[f]These definitions are taken from the Federal Energy Administration, *Monthly Energy Review* (March 1975), pp. 60-62.

the current $5.25 a barrel. Since the exploration and drilling activity has already taken place, it simply has to be lifted from the reservoir. The problem occurs because the definition of old oil does not take into account the fact that there are natural declines in production rates from old fields. Thus, while in fact the production rate of old oil is declining from a given well, the definition, because it is based on a fixed historical period, remains constant. As a result any additional production that might occur through expensive secondary and tertiary recovery techniques would be classified as old oil and could be sold only at $5.25 a barrel. Since these techniques are expensive, this does not provide an adequate incentive.

The second problem, an increased incentive to import, results not from the program to control prices but rather from a complementary program designed to reduce the disparity in petroleum costs among different regions of the country.[g] Since old oil is cheaper than new oil or imported oil, if regional petroleum costs are to be equalized different regions have to have equal access to the low-cost old oil. This is accomplished through the crude oil equalization program administered by the Federal Energy Administration. Under this program entitlements to the cheaper old oil are rationed to refineries on the basis of total oil use. All refiners get an entitlement to 40 percent of a barrel of low cost crude for every barrel they use. These entitlements sell for about $3 a piece, which means that the true cost of a barrel of imported oil is actually $3 less than the market cost. Because of this pairing of imports with old oil, there is, in effect, a subsidy paid to importers of oil.

What is the appropriate policy response? Secondary and tertiary recovery methods are extremely desirable sources of oil for environmental reasons. These occur in established fields where the local areas have already adapted to the presence of oil production. The schools are built, the roads and pipelines are in place, and the skilled labor is already there. These techniques make known fields yield more oil. For these reasons plenty of incentive should be supplied to insure these techniques are applied.

Yet, it is equally clear that for truly old oil a $5.25 price is quite adequate. The solution, therefore, is to allow the definition of old oil to be based on declining production rates. Anything above those rates would, as now, be classified as new oil and hence would receive higher prices. The political compromise would take place over the rate of decline. More rapid decline rates would yield more rapid increases in average petroleum prices.

This represents a compromise solution in that the ultimate objective of decontrol is established in line with administration objectives but the phasing in would likely be slower than the administration would like it. It is quite likely that the crude oil equalization program would continue during this period of

---

[g]The nature of this program and its effects on the incentive to import are described in Michael Canes, "The Effect of Old Oil Price Decontrol on Product Prices," mimeo, May 9, 1975.

transition out of decontrol for the simple reason that it is necessary to protect both independent refiners, who have less natural access to low-cost oil than the majors, and the people in the Northeast, who depend heavily on imported oil. This political necessity would unfortunately retain the perverse incentive to import but this incentive would diminish over time and should be viewed as one of the costs of compromise.

The next major issue is the deregulation of natural gas. This is a tricky issue and it is important to realize that many of the strong arguments of a few years ago have lost some of their validity, since rather substantial increases in the controlled price have taken place since then. In 1972 the average initial rates paid by interstate pipeline companies for natural gas under new long-term contracts was about 24.5 cents per mcf.[5] By 1974 this price had risen to 43.9 cents per mcf,[6] with the ceiling currently set at about 50 cents per mcf.[7]

Natural gas is politically a very difficult fuel to deal with because its unregulated price would soar above the costs of production for most forms of natural gas and certainly above the average cost of production. Huge profits would result because the price of natural gas would be determined primarily by the cost of its closest substitute, which, since electrical power plants are the largest consumer of natural gas, is residual fuel oil. As a result the unregulated field price could be expected to rise to about $1.40 per mcf or almost three times the current price![h]

What would be gained by deregulating? In addition to correcting the misallocation of resources between interstate and intrastate markets, which was discussed in Chapter 2, deregulation would restrain demand growth and increase supply. The crucial variable is the supply responsiveness and we do not know very much about that, particularly at prices above $1.00 per mcf.

Once again, in an attempt to provide a complementary political and economic response to this problem it is imperative that the goal of deregulation be accepted and the compromise worked out over the time phasing of deregulation. The price to consumers of natural gas should be raised to its market equilibrium level as soon as possible so the process of substituting other fuels, where efficient, can begin to take place soon. This will also solve the regional disparities, which arose because only interstate gas was controlled.

Since this will create windfall profits, this proposal should be accompanied by a windfall-profits tax. This tax plan should have a specific schedule for phasing the tax out so that all parties can plan on future resource allocations under stipulated ground rules. By establishing a firm decision rule, which is clear to all participants, planning can take place by both natural gas users and natural gas producers. If the world price for crude should fall, then the problem would become less severe and the tax could be lowered or eliminated.

---

[h]This estimate was provided to me by Hill Huntington at the Federal Energy Administration. It takes into account the BTU content of natural gas *vis-à-vis* residual fuel oil plus the distribution and transmission costs involved.

Immediate decontrol of oil prices and deregulation of natural gas prices is politically unlikely. Therefore, one is forced to establish priorities and to enter compromises. These compromises should reflect the fact that the natural gas problem is currently more serious and that natural gas is environmentally a superior fuel. Therefore, if there is a constraint on the total amount of upward pressure on prices, as there seems to be, then rapid natural gas price rises should be given priority over oil prices.

The third major area of controversy is the decision over whether to accelerate the development of our domestic oil reserves. Ignoring environmental costs and the future scarcity costs incurred by drawing down domestic resources at a faster rate, the answer is unambiguously affirmative. A substantial proportion of our domestic oil reserves appear to be economically viable at prices substantially under the current price of imported oil.

Would the inclusion of these environmental and society costs change the decision? The answer is that we simply do not know. In part, we do not know because the amount of oil involved is not known with any kind of precision. In part, we can never know with certainty because environmental costs are so amorphous and quantification attempts are so primitive, and because the timing and availability of potential substitutes is not yet at all clear. This remains largely a political decision, but that decision is difficult to make without specific knowledge about the size of each of the fields involved and about the technical feasibility and economic viability of fuel substitutes for oil and natural gas.

As has been shown above with a strategic reserve in place, the need to accelerate the development of domestic reserves is not that pressing. A definite distinction should be drawn between the application of secondary and tertiary recovery techniques to existing fields and the development of the various Outer Continental Shelf areas. With the decontrol plan suggested above secondary and tertiary recovery could be expected to provide as much as 4.7 million barrels a day by 1985 if the price remains high.[8] This would lessen the need for an immediate heavy development of the Outer Continental Shelf.

The Outer Continental Shelf resources should be intensively explored to find out what is there, but the decision about what to develop, and when to develop it should await this information. The current system of leaving the timing and field selection up to the oil companies can lead to suboptimal decisions. This is appropriately a political choice.

The acceleration of the construction of nuclear power plants appears absolutely unnecessary. The two main arguments for accelerating their development are: (1) a vast expansion of generating capacity will be needed if blackouts are not to occur in the 1980s, and (2) the nuclear plants will replace oil fired plants and this will diminish the import vulnerability problem. Both arguments appear, on the basis of preliminary information generated by the Federal Energy Administration, to be factually incorrect. The demand argument is based on assumed very high overall growth rates in electricity demand, in spite of high prices, and in the more rapid growth of peak-hour to off-peak-hour electricity consumption. The latter problem, if real, is acute because the level of peak-hour consumption determines the capacity needs. This implies that an increasing amount of capacity is planned to be inactive during the off-peak period, since electricity storage on a large scale is not yet practical.

The demand arguments used to substantiate a need for a massive program of nuclear development are suspect because recent econometric work at FEA[9] indicates that the demand growth in electricity is likely to be substantially below industry forecasts because of the recent large increases in electricity prices. In addition, the adoption of peak-load pricing[i] or other load management techniques, which are already being adopted in many places, should limit the growth of peak-hour to off-peak-hour consumption.

The second argument, that the rapid adoption of nuclear plants will lower import vulnerability is, according to the *Project Independence Report*, also in error.[10] The greatest substitutability is between coal and nuclear plants rather than between oil and nuclear plants. For this reason the speed with which nuclear plants are constructed will have very little to do with import vulnerability.

What is to be gained by moving more slowly into nuclear power? Most importantly we will be able to allocate more time to solving the potentially large problems associated with storing the highly radioactive spent wastes for literally thousands of years. These problems are so important that we cannot proceed into a large-scale nuclear development program until they have been solved. It is not enough to proceed on the assumption that when the need for a solution appears the solution will soon follow.

There will continue to be some growth in electricity growth, of course, and for every nuclear plant not constructed to meet this growth, a coal plant will have to be constructed. The environmental dimension of coal cannot be overlooked because it is a major source of air pollution in the form of sulphur oxides. Nonetheless, when the choice is between some additional sulphur oxide pollution in the short run, until new sulphur eliminating technologies are perfected, and radioactive wastes, for which there is no known adequate disposal method, the choice, to me at least, is quite clear.

The thrust of this chapter is to suggest that workable compromises, which can preserve the responsiveness of the economic system while meeting some of the political objections being raised in Congress, are possible for energy. Whether the political institutions can arrive at these compromises or others with similar properties remains to be seen.

---

[i]Peak load pricing involves charging higher rates for electricity consumed during peak hours than for electricity consumed during off-peak hours. This encourages the substitution of off-peak for peak consumption and diminishes the need for expanding capacity.

Notes

1. Davis B. Bobrow and Robert T. Kudrle, "Theory, Policy and Resource Cartels: The Case of OPEC" (Paper delivered to The American Political Science Association, Chicago, Illinois, August 29-September 2, 1974).

2. Ibid., p. 57.

3. Joseph A. Yager and Eleanor B. Steinberg, *Energy and U.S. Foreign Policy* (Washington: Brookings Institution, 1974), table 13-7, p. 252.

4. Federal Energy Administration, *Project Independence Report* (Washington: U.S. Government Printing Office, 1974), table VII-3, p. 358.

5. Federal Power Commission News Release, 5 May 1975, No. 21367.

6. Ibid.

7. Federal Power Commission, *A Preliminary Evaluation of the Cost of Natural Gas Deregulation*, January 1975, p. 17.

8. Federal Energy Administration, *Project Independence Report*, p. 83.

9. Federal Energy Administration, *The Preliminary Report of the Electricity Task Force*, 30 June 1975, p. 28.

10. Federal Energy Administration, *Project Independence Report*, p. 8.

10 Lessons

As of October 31, 1975 only a few scattered pieces of energy legislation had become law. Aside from the reorganization acts (creating the Energy Research and Development Administration—ERDA and the Federal Energy Administration—FEA) which were not particularly volatile issues and the acts passed during the height of the crisis, not a single major energy act had become law. The recognition of this statement is less important, however, than an analysis of its underlying causes and a forecast of their potential long-run implications.

An identification of the forces that led to this stalemate provides an important point of departure. It permits us to generalize from this experience and to discuss the likely response of the policy-making process to energy problems in the future as well as to problems that share some of the important characteristics of the energy problem. Is there cause for pessimism about the ability of political institutions to cope with energy or, perhaps, other depletable resource problems? What were the main impediments to a responsive policy process? How serious are these impediments?

This analysis of the underlying causes also provides a basis for a discussion of reforming the institutions or procedures. Are these impediments inevitable or can they be circumvented through carefully designed reform measures without sacrificing other important political goals, such as the movement toward increased democratization of the congressional organization and procedures?

Responsiveness of the Policy Process

In assessing the responsiveness of the policy-making process two issues are paramount: (1) the ability of the process to provide a comprehensive and coherent policy framework, and (2) the ability to design a policy framework which is flexible enough to cope with uncertainties and unanticipated outcomes. Both are important dimensions. The presence of a policy framework defines the ground rules under which the economy operates. It reduces uncertainty and guides the decentralized decision process that characterizes the market system. Conversely, the absence of a policy framework increases uncertainty. Procrastination in the policy process leads to procrastination in the private sector. Important decisions are deferred until the policy guidelines are clear.

Flexibility in the resulting policy framework is also important, especially when the process itself is not sufficiently flexible to undertake systematic policy

141

reevaluations every few years. The policy process is so cumbersome and requires such a huge expenditure of effort on the part of participants that it provides an enormous incentive for permanent or quasi-permanent legislation. The policy agenda is so filled with new business that it is difficult to squeeze in reevaluations of past decisions. Because of this, it is especially important to legislate a policy framework that can adapt to changing circumstances.

In Chapter 1 some a priori reasons for concern about the ability of the policy process to provide a comprehensive, coherent, and flexible policy framework were presented. We can now review those concerns and assess their validity in the light of the Project Independence experience. The concerns generally fell into three categories: the complexity of the issues coupled with inadequate information channels to bring the available information to bear on these issues, the inherent and visible value conflicts that would make compromise a more illusive vehicle for producing policy, and institutional rigidities that would preclude effective management of the policy framework.

There are two levels of issue complexity. The first level is the complexity associated with forecasting the consequences of particular proposals and relating those consequences to the aggregate objectives for that program. The second level is the legal complexity of translating this program concept into specific legislation, with all responsibilities and instruments clearly articulated.

The first level of complexity was not, in retrospect, a major impediment. A model was developed and it was used in the policy process. The preconditions for this successful integration are, in part, replicable and, in part, perhaps not. The replicable characteristics were the design of the system itself, and an unswerving respect for deadlines. The system design provided information without infringing on the prerogatives of decision makers by concentrating on estimating multidimensional consequences of policies rather than providing a unidimensional ranking of policies. It also permitted key participants in the policy process to examine the likely consequences of policy packages of particular appeal to them before they became publicly committed to a particular program. Other factors in the success may or may not be replicable because they depended upon the personalities of some of the principal participants and their relationships with each other. The availability of a translator, who thoroughly understood the strengths and weaknesses of the model, the availability of informal channels of communication within the executive branch and between the executive branch and Congress, and the willingness on the part of the principles to be guided, at least in part, by the analysis are cases in point. These are transient phenomena, embodied in people, not institutions. While it is quite likely that there are many possible combinations of people that could produce similar results in future policy situations, there is nothing automatic about insuring that these people will occupy the appropriate positions when the need arises.

The second level of complexity, translating a broad program design into specific legislation, is more problematical. In part this is due to the intense conflicts of values, which were, and are, an inherent part of the energy policy process. The executive branch, as expected, was, in the formulation of its own plan, somewhat less susceptible to these value conflicts than was Congress. There were several reasons for this. In the first place the executive branch had a fairly uniform ideology that it applied to the problem. The overriding objectives within the executive branch were minimizing the role of government and reducing import vulnerability. The second contributing factor was that the President could be somewhat isolated from the adverse political consequences of a program that would increase unemployment and inflation somewhat by virtue of the countervailing political gains accrued from his decisiveness in dealing with the energy issue after a long period of inactivity. Since the announcement of his energy and economic program reversed a decline in the popularity of President Ford, which had persisted since he assumed the office, it certainly appears that decisiveness has political merit quite apart from the substance of the decision.[1]

Congress, of course, is the most susceptible part of the policy process to value conflicts. Indeed, as the most representative branch of government, the Founding Fathers assigned Congress the role of being the central arbiter of these conflicts. During Project Independence, this proved to be an exceedingly difficult task.

The process of coalition formation proved very difficult because there were so many diverse coalitions involved. One could not simply appeal to members of Congress along party or ideological lines; the issues had too many regional implications and were too visible to their constituencies.

In assessing this process of coalition formation and its effect on the shape of legislation it is possible to isolate some key tendencies of the policy process and to examine the implications of these tendencies should they persevere. Two such tendencies stand out as being of potentially large importance. The first is the tendency of Congress, in the face of increasingly complex issues, to allocate more discretionary authority to the executive branch without the accompanying authority to carry out this responsibility. The second tendency is for successful coalition formation in Congress, particularly in the House of Representatives, to require strong restraints on price increases. Each of these tendencies is illustrated below by reference to specific occurrences during the energy policy process and then their implications are discussed.

The tendency of Congress to prefer discretion to rules can be illustrated by the fate of the strip mining legislation. The control of strip mining is generally agreed to be desirable; the debate hinges on the degree of control. Congress, as it wrote the strip mining legislation, realized that it could never hope to specify all the standards and procedures to be followed in the legislation. Therefore, it empowered the Secretary of the Interior to draw up the procedures that would fulfill the general principles of the act. His interpretation of the act would be subject to judicial review.

One of the grounds apparently given by the administration for vetoing the act was that several of the principles against which the Interior Secretary's rules would be measured were sufficiently ambiguous to make judicial resolution virtually a certainty.[2] This would inevitably cause more delay and more uncertainty.

The purpose in citing this case is to point out a potentially serious impediment in the policy process, which arises when the problem to be addressed is technically complex. Since Congress cannot hope to specify completely the policies, it must either allocate a good deal of discretionary authority to the executive branch or it must specify decision rules that are fairly specific, yet flexible enough to be satisfactory under a variety of contingencies. The discretion option has a particular appeal unrelated to its inherent desirability as a mechanism to guarantee flexibility; it provides a bureaucratic scapegoat if the program fails. This provides a direct political analogue to a concept in economics known as externalities. The basic thrust of the concept is that when some of the costs and/or the benefits of a particular decision fall on someone other than the decision maker they will not be considered correctly in making the decision. As a result it is likely that an inefficient decision will be made.

The application of this concept in this context would imply that by stating fairly general purposes in a piece of legislation and requiring the executive branch to design procedures for fulfilling these purposes, Congress can duck the politically difficult issues. If the devised procedures stir up a storm of controversy, Congress can intervene after the fact. If the devised procedures fail to achieve the desired objectives, the bureaucracy can be blamed for designing a poor system.

If this is an accurate characterization of the congressional choice between rules and discretionary authority, it suggests that the preference for discretion may result not because these are merely technical details best left to experts, but rather because the discretion option allows Congress to create the appearance of solving problems without incurring the political wrath of those who would oppose the specifics of the program. When this characterization is coupled with the inevitability of judicial review and the tendency of the court system to take these vague legislative preambles quite seriously, the result is, at best, further delay and, at worst, a series of court instituted procedures that are counterproductive.[a]

The fate of the Energy Conservation and Conversion Act of 1975, as it passed through the House of Representatives, is another interesting example of choices between rules and discretion as well as an example of the reluctance of Congress to adopt rules that would result in higher consumer prices. In that bill, the Ways and Means Committee had mixed the allocation of discretionary authority with

[a]This tendency of the judiciary to play an increasingly large role in policy formation has also been suggested by Nathan Glazer, who views this tendency with some alarm. See Nathan Glazer, "Towards an Imperial Judiciary," *The Public Interest* (Fall 1975): 104-23.

the use of some clearly specified, dynamic, contingent, decision rules. The final bill had eliminated most of the price-based decision rules and increased the amount of discretionary responsibility in the executive branch.

Title I of that act specifies the maximum amount of petroleum imports to be allowed into the country. These limits were to prevail unless the President determined that some other import level was in the national interest. In this case he could modify the quantitative limits up or down within a range specified by Congress (a million barrels a day until 1977). After modifications on the House floor, the procedure remained intact, but the maximum import levels had been raised.

Title II was in some ways the most interesting section of the act because it contained specific, flexible rules, which limited the authority of the implementing agency, in this case, the Federal Energy Administration. In Title II, a straight gasoline tax of 3 cents a gallon was supplemented by a more stringent tax, up to 23 cents per gallon, which would go into effect only under certain clearly defined conditions. The FEA administrator was empowered to determine whether gasoline consumption in the preceding year exceeded gasoline consumption in 1973 and, if so, by how much. Once this determination was made the law specified the amount of the increased tax to be imposed. This was a useful way for Congress to be clear in its intent while providing some flexibility in the policy. The fact that the taxes were contingent on a specified event meant that, if, because of private actions or because of other policies, this event would be achieved even without the taxes, the taxes would not be imposed. Unfortunately, this provision did not survive the floor debate. A noncontingent tax on business use of petroleum, petroleum products and natural gas did survive the floor debate, presumably because it would be much less visible to voters.

Similarly the excise taxes on new automobiles getting low gas mileage were replaced with a system of civil penalties for automobile manufacturers that did not meet certain fleet standards for gasoline mileage. In this case a specific price-based rule, the excise tax, was replaced by a civil penalty, which could be removed at the President's discretion and which could be more easily challenged in the courts. Apparently these characteristics were important in the congressional substitution of the civil penalties for the excise taxes.[3]

The reasons for the deletion of these visible taxes are many and complex. It is therefore not possible to assess the extent to which their deletion was due to the peculiar circumstances at the time (e.g., high rates of inflation and unemployment) and the extent to which it was a general rejection of a particular policy instrument, taxes, imposed not primarily for raising revenue, but rather for modifying behavior. There is a clear precedent for this latter interpretation because a very similar reaction by Congress occurred when proposals to tax effluent were eliminated in favor of the adoption of an allocation of discretionary power to the Environmental Protection Agency to set effluent standards for discharges into the nation's water resources.[4] In both cases, it was a specific rejection of a price-based rule for a discretionary policy.

There are other examples that illustrate congressional reluctance to be a party to higher prices. It was stated in the previous chapter that potentially workable compromises exist if Congress will accept the long-run goal of releasing prices from government control (e.g., domestic oil prices and natural gas prices) in return for a system that phases these changes in rather slowly. This would allow the economic system to do what it does well, retard the growth in energy demand and stimulate the discovery of new energy sources. If Congress refuses to accept this kind of compromise, as they have so far, the consequences of a political impasse would be much more severe. Political nonresponsiveness, in this case, would contribute to a lack of responsiveness in the economic system; the automatic character of this economic response, the basis for the strong criticism of the *Limits to Growth* thesis, would have been prevented. Without this adaptability it is less possible to be quite so sanguine about the ability of our social institutions to deal with problems of increasing scarcity.

The final source of concern about the ability of the policy process to respond, which was suggested in Chapter 1, was that the institutional structure within which the process operated was inappropriate for handling the energy problem and was resistant to change. The fragmentation of jurisdiction over energy matters in both branches of government was a well recognized fact. The executive branch proved more malleable in adapting to meet the needs of designing a comprehensive program. The formation of the Energy Resources Council was, when judged solely in terms of its ability to produce an energy policy framework, an effective step. On other grounds, such as its failure to represent environmental and consumer interests adequately and its inability to resolve the Kissinger-Simon disputes completely, it was less successful.[b]

The fragmented jurisdiction over energy matters within Congress was more persistent and more important. Attempts to provide a structure more congenial with the basic needs of energy policy formation were unsuccessful. This fragmented jurisdiction has caused a variety of problems. First, it led to some jurisdictional disputes, which have delayed important legislation. These disputes occurred over the bill establishing a windfall-profits tax by the Senate Finance Committee, which usurped a traditional Ways and Means Committee prerogative initiating all tax legislation. They also occurred over the various bills authorizing opening the Naval Petroleum Reserves to supply oil for the civilian economy.

Fragmented jurisdictions also cause problems in negotiation between Congress and the President. There is no power center or no central spokesman with which to negotiate. This situation is a bit like asking the management of a business to negotiate with the entire labor union. In the past these problems were somewhat less intense because there were informal power centers. Powerful

[b]Environmental concerns were represented on the Council by the heads of the Council on Environmental Quality and the Environmental Protection Agency. Consumer interests were represented by the President's special assistant on consumer affairs. The rather small role played by these three members is illustrated by the fact that not a single one was at the Vail meetings where the final package was put together.

committee chairmen could produce results simply by throwing their prestige behind a bill, or, if that failed to muster enough votes, by dispensing various rewards and penalties to the members of their committees. Earlier in our history the leadership function was performed by the party caucus. A consensus could be developed on issues because voting with the party was an enforced tradition.

These forces have largely disappeared now and in their place has evolved an increasingly democratic representative organization. One of the effects of this process is that Congress is a less effective organization for getting things done. This poses a fundamental dilemma because it implies that increased democratization, as it is currently being implemented, diminishes the ability of Congress to act effectively and rapidly. To the extent that these two goals are inevitably contradictory the process of institutional reform will require a careful consideration of the trade-off between the two.

The rules and procedures of Congress are very supportive of the status quo. It is much easier to stop legislation than to pass it. Coalitions consisting of a minority of congressmen can block action on legislation either through a filibuster in the Senate or killing a bill in committee in the House. The Presidential veto is another source of support for the status quo. This was used to defeat the strip mining bill and has been used repeatedly on the attempts by Congress to resolve the oil decontrol issue.

In retrospect the only major energy legislation passed by Congress that has survived the veto were the various executive reorganization bills (creating FEA and ERDA), which were not particularly politically volatile, the bills passed during the height of the embargo in the prevailing crisis atmosphere and the tax-cut bill with the repeal of depletion allowance tacked on as an amendment. The depletion allowance repeal was a successful use of a rider by Congress because they knew it would be political suicide for the President to veto the tax-cut bill. Attaching riders can be an effective way to compromise by pairing issues, but it only works when certain preconditions are fulfilled and these preconditions are rarely present. Not only is a veto usually an effective cudgel to get Congress to unbundle the combined package, but, because of the fragmented jurisdictions within Congress, the effective use of riders would frequently require the cooperation of two committees, a rare feat.

Possibilities for Reform

This examination of energy policies has raised a number of interesting hypotheses about the nature of the political system and its ability to respond to certain stimuli. It also suggests some areas where fruitful investigation into institutional reform might be undertaken.

No attempt is made in this book to outline specific reform proposals. That would be a rather presumptuous way to end a book that is concerned mainly with an investigation of a particular policy event. Yet, this event is so rich in information about impediments and strengths of the policy process in establishing a rule of law that it does suggest some areas where specialists in institutional reform might fruitfully concentrate some effort.

The first such area concerns the ability of Congress to use effectively the comprehensive planning models that are now being developed. These models have an important role to play in the policy process, but the current institutional setup within Congress makes their use difficult.

Planning models provide two very important services to decision makers: (1) they can forecast the likely outcomes of a policy in an internally consistent and historically consistent manner, and (2) they can estimate the cumulative impacts of multiple policies in conjunction with one another. These are particularly valuable services when the policy framework is constructed sequentially, one policy after another, and when the policies are drawn up by different committees. The cumulative benefits to be derived from these programs in terms of their effects on energy production, energy demand, or imports are rarely additive. Some mechanism is needed to keep track of the cumulative aggregate benefits to be derived from a sequence of bills, and this information has to be made generally available to the members of Congress.

The creation of a single committee to handle all energy legislation is politically unrealistic and, after all, would fragment other responsibilities of Congress (e.g., the responsibility of the Ways and Means Committee in the House for handling all tax legislation). The evidence is quite clear that any realistic reform measures will have to deal with Congress pretty much as it is currently structured.

One alternative, which appears worthy of additional consideration, is to broaden the role currently being played by the Congressional Budget Office. On budget matters, this committee serves as a central repository for handling large-scale budget modeling. This centralization of authority explicitly recognizes the economies of scale involved in running these kinds of models as well as in interpreting the results and it recognizes the necessity for an overview of the component parts of the process.

If the responsibility of the Congressional Budget Office could be expanded so that it would become the central repository for all planning models, not only economic stabilization planning models, such a move would appear to be quite beneficial. It would not only provide Congress with better information, and therefore better control, in establishing an energy policy framework, it would also provide an analytical counterpoint to the analysis provided by the administration. This would inevitably lead to a better dialogue between the administration analysts and congressional analysts on techniques, assumptions, etc. It is hard to believe that this would not improve the understanding by both Congress and the administration of the implications of their actions.

The current system by contrast has the staff of each committee struggling with the modeling system. Not only does this generally overextend these staffs, it also makes an ineffective use of the administration analysts who, as soon as they have educated one staff in the nature of the modeling system and its use, have to go through the same process all over again with the next staff. A single point of contact would not only breed better understanding of the system, it would also eliminate a lot of wasteful duplication of effort.

The next major area for investigation is the congressional use of rules versus discretion in formulating policy guidelines. As was pointed out above, the evidence that is presented by the energy policy process is consistent with the hypothesis that the Congress is biased away from the price based rules and toward nonprice discretionary systems. If this hypothesis is correct, it implies that the set of politically feasible policy instruments does not contain those instruments that are most likely to retain the natural responsiveness of the economic system in managing the rate of depletion of domestic exhaustible resources. Furthermore, it is likely that the seriousness of this bias, if it exists, will increase over time.

To illustrate this point, consider the effect of a continuation of the policy to hold down artificially the price of natural gas. Natural gas is a prime substitute for residual fuel oil. In a market situation in which there is not enough natural gas to substitute for the residual the price of natural gas will be determined by the price of residual. This price is substantially above production costs for currently produced natural gas. As a result, in a free market, producers would be selling their natural gas at a price far above costs. In the longer run, of course, these higher prices call forth additional supplies with higher costs of production. Yet, it is quite likely these will coexist with the low-cost sources for some time and these low-cost sources will still be receiving large windfall profits. This situation will occur repeatedly in conditions of increasing scarcity and will be particularly acute when the introduction of new sources (e.g., synthetic gas produced from coal) leads to significantly higher production costs.

The clear tendency of Congress is to react to this situation by holding prices down to something near the average cost of production. In conditions of increasing scarcity this makes the adjustment process more difficult by failing to send out the appropriate price signals to users to get them to substitute more abundant sources for the scarce ones and to conserve; it also fails to provide any incentive for producers to bring the high-cost sources into production.[c] The result is that what is currently available is sold quickly and not replaced by new discoveries.

---

[c]The relationship of prices to the rate of depletion of our domestic exhaustible resources is described in some detail in Robert Solow, "The Economics of Resources or the Resources of Economics," *The American Economic Review*, LXIV (May 1974): 1-14. A more technical treatment of the underlying theory can be found in Orris C. Herfindahl and Allen V. Kneese, *Economic Theory of Natural Resources* (Columbus: Charles E. Merrill Publishing Co., 1974) and in the special symposium on this subject published as a special 1974 issue of *The Review of Economic Studies.*

This problem is currently particularly acute for natural gas because the price of its closest substitute is about three times the current regulated price for natural gas. Since this price is artificially high (i.e., the high price of its closest substitute, residual oil, does not truly reflect world scarcity), it is unlikely that other resources will have such a large divergence between the costs of production for old sources and new sources. Yet, such differences will occur. When they do, if the political system insists on maintaining low-administered prices, the ability of the economic system to respond will be diminished. To the extent that Congress uses a comparison of price with the average cost of production as an equity standard instead of a more reasonable real income standard, relying on price increases coupled with rebated windfall-profit taxes, the political system will increase nonadaptability in the economic system. And to the extent that Congress fails to stimulate energy conservation through the judicious use of taxes, the rate of depletion of our domestic exhaustible resources will be higher than prudence would dictate.

The final area for research on reform is to investigate ways to use the status quo orientation of Congress to increase the flexibility of the political and economic system. This is not as much of a contradiction as it might appear at first. For example, policies that specifically retard the adaptability of the economic system could be enacted for specific periods of time. Continuation of such programs would require specific renewing legislation.

This approach would have several merits. It would automatically place any such policy on the policy agenda when the original law expired. Second, it would put a deadline on congressional action. Since Congress seems to move more rapidly in the face of deadlines, this would make some action more likely. Finally, it would tend to preserve the adaptability of the economic system in the presence of a political impasse.

Concluding Comments

Project Independence was, and is, a somewhat unique experience in the American political process. In this book, I have attempted to provide an interpretive history of the formative stages of that experience with an emphasis on the role of comprehensive planning models.

This experience seems to indicate many things. Comprehensive planning models can provide a valuable input to the policy process, but the existing information channels for their use tend to be informal and less than totally effective, particularly in Congress. An examination of the forces leading to the current lack of policy responsiveness reveals that some of them are likely to persist unless specific action is taken to circumvent them. An examination of the policy instruments chosen by Congress to implement its will suggests that when Congress responds, it may well do so in ways that are inconsistent with maintaining a prudent rate of depletion of our domestic exhaustible resources. It would be wrong to interpret these findings as suggesting that the rule of law, as it is expected to evolve, will validate the *Limits to Growth* thesis by building an increasing degree of nonadaptation into the economic system. Yet, there are enough pessimistic signs to warrant caution in accepting the view that the political and economic system in which we live is automatically adaptive. The search for ways to maintain or increase adaptability while respecting the need for increased citizen participation in, and democratization of, the policy process is a valid and important exercise. It may prove to be the ultimate test for democracy.

Notes

1. *The Gallup Opinion Index*, June 1975, p. 3.

2. Arthur J. Magida, "Environment Report/New Strip Mine Bill Ignores Ford's Suggestions," *National Journal Reports*, 8 May 1975, p. 370. This point was also made in the postveto hearings held on June 3, 1975.

3. *The New York Times*, 13 June 1975, p. 44.

4. Allen V. Kneese and Charles L. Schultze, *Pollution, Prices and Public Policy* (Washington: Brookings Institution, 1975), pp. 97-98.

Appendixes

Appendix A:
Methodologies

The assessments in Chapter 9 require a quantification of the costs associated with implementing a variety of policy packages and of the benefits to be derived from their implementation. The benefits are defined as the reductions in the expected costs of future oil embargoes. In this appendix the assumptions and procedures used to provide this quantification are documented.

The methodology involves a blending of now familiar microeconomic constructs[1] with the analytical power of the Federal Energy Administration (FEA) energy modeling system. The latter is used to quantify the effects of various policy packages on key variables such as energy consumption, domestic production and import levels. The former is used to assess the costs and benefits of undertaking such a project.

The cost and benefit calculations are performed for each of the 11 years in the 1975-85 period. In any year, by assumption, any policy could lead to four different possible outcomes. Which of these outcomes will prevail is assumed to be completely out of the control of the United States government. These outcomes are determined by Organization of Petroleum Exporting Countries (OPEC) pricing strategies and OPEC embargo strategies. For this analysis it is assumed that OPEC chooses to follow one of two pricing strategies: either it maintains continued high prices (more precisely $11 a barrel in 1975 prices) or it drops the price to $7 a barrel (in 1975 prices) in 1978 and holds it there through 1985. OPEC also can choose whether or not to initiate one or more embargoes in the next 11 years. The combination of these contingencies provides the four possible outcomes for each year.

Three categories of cost are computed: the cost of an embargo or disruption, the cost of a strategic reserve to provide emergency replacement oil supplies, and the cost of enforced conservation. Each of these is discussed in turn, followed by a discussion of the procedures used to reduce all these cost elements to an unidimensional index, the present value of expected cost.

The Cost of an Embargo

Any forecast of the cost of future embargoes must deal with three issues: the size of the embargo, the duration of the embargo, and the year in which it is presumed to strike. The latter two are dealt with in this book by assumption. The duration is assumed to be roughly the same as the last embargo or 180 days. Since there is no apparent reason to believe that an embargo would be more likely in one particular year than another it is assumed to be equally likely in any one of the 11 years.

155

The expected size of the embargo is calculated using the information provided by the *Project Independence Report* about the proportion of imports, as a function of the import level, which could be expected to come from embargo prone sources. It turns out that the results of this analysis can be adequately approximated by a simple linear formula:

$$\overline{Y}_{tj} = a + b\overline{I}_{tj} \qquad t = 1, \ldots, 11; \quad j = 1,2 \qquad (A.1)$$

where $\overline{I}_{tj}$ is the level of petroleum and petroleum product imports in millions of barrels per day for year t and OPEC pricing strategy j and $\overline{Y}_{tj}$ is the expected level of insecure imports. The "$-$" notation is used to denote variables referring to a preembargo equilibrium. The total import levels are determined for each year, each price scenario, and each policy package by the FEA modeling system. The coefficients $a = -2.7$ and $b = +0.76$ were chosen because the resulting equation represented a good approximation to both current and expected future import situations.[a] The embargo is assumed to affect all embargo prone sources.

The next step is to calculate the during embargo equilibrium levels for prices and quantities. The during-embargo consumption level is assumed to be simply the preembargo consumption level minus the effective embargo. The effective embargo is simply the embargo-prone imports minus the amount of storage used during the period of the embargo. Symbolically,

$$\hat{Q}_{tj} = \overline{Q}_{tj} + 2.7 - 0.76\overline{Y}_{tj} + S_{tj} \qquad t = 1, \ldots, 11; \quad j = 1,2 \qquad (A.2)$$

where $\hat{Q}_{tj}$ is the consumption level during the embargo and S_{tj} is the amount of the strategic reserve used during the embargo. The "$\hat{}$" notation refers to variables characterizing the during embargo equilibrium.

The next step is to calculate the implied price rise that would equate the domestic demand with the remaining supplies. Since, in the short run, the supply elasticity can be expected to be zero, this price will depend solely on the nature of the demand curve.

This curve can be specified with two pieces of information: the functional form of the demand curve and the values of its parameters. The functional form is chosen to yield a constant short-run price elasticity of demand. When the estimated elasticity used by the FEA modeling system (0.11) is substituted for the elasticity parameter, the remaining parameter can then be found by solving (A.3) for A_{tj}.

$$A_{tj} = \overline{Q}_{tj} \cdot \overline{P}_{tj}^{a} \qquad t = 1, \ldots, 11; \quad j = 1,2 \qquad (A.3)$$

---

[a]This curve, for example, passes through the current situation of about a 30 percent dependency for 6.5 million barrels a day and a projected future dependency of 6.2 million barrels a day if total imports rise to 12.0 million barrels a day. See Federal Energy Administration, *Project Independence Report* (November 1974), p. 363.

A_{tj} is the unknown parameter, a is the estimated elasticity of demand, $\overline{Q}$ and $\overline{P}$ are points on the known long run demand curve. This formula can then be used to find the during embargo implied price.

$$\hat{P}_{tj} = (A_{tj}/\hat{Q}_{tj})^{1/a} \qquad t = 1, \ldots, 11; \; j = 1,2 \qquad \text{(A.4)}$$

The next step is to use this information to compute a total cost to energy users of the shortfall. Following conventional practice this is defined as the area to the left of the derived demand curve between the initial and final prices, which can be found by integration.

$$C_{tj} = \int_{\overline{P}_{tj}}^{\hat{P}_{tj}} A_{tj} P_{tj}^{-a} \, dP_{tj} \qquad t = 1, \ldots, 11; \; j = 1,2 \qquad \text{(A.5)}$$

where $\hat{P}_{tj}$ is the during embargo price and $\overline{P}_{tj}$ is the preembargo price. The solution to this integral can be shown to be

$$C_{tj} = \frac{A_{tj}}{1 - a} \left[\hat{P}_{tj}^{1-a} - \overline{P}_{tj}^{1-a} \right] \qquad t = 1, \ldots, 11; \; j = 1,2 \qquad \text{(A.6)}$$

While this is an accurate representation of the cost of the embargo to energy users, it is not an accurate representation of the costs to the nation as a whole. It is an overestimate because it includes two income transfers: the windfall profits earned by the domestic oil industry due to higher prices and the revenue accrued by the government by selling part of the strategic reserve.

Since the size of this transfer (F_{tj}) can be represented symbolically as

$$F_{tj} = (\hat{P}_{tj} - \overline{P}_{tj})(S_{tj} + \overline{Q}_{tj} - \overline{I}_{tj}) \qquad t = 1, \ldots, 11; \; j = 1,2 \qquad \text{(A.7)}$$

then the total cost of the embargo, net of transfers is calculated as:

$$G_{tj} = C_{tj} - F_{tj} = \frac{A_{tj}}{1 - a} \left[\hat{P}_{tj}^{1-a} - \overline{P}_{tj}^{1-a} \right]$$

$$- (\hat{P}_{tj} - \overline{P}_{tj})(S_{tj} + \overline{Q}_{tj} - \overline{I}_{tj}) \qquad t = 1, \ldots 11; \; j = 1,2 \qquad \text{(A.8)}$$

As a check on the validity of this model it was used to calculate the cost of the last embargo and the implied price. The implied price was \$11.65 a barrel, which conforms well to published figures,[2] and the estimated cost was $12.3 billion.[b] This is in accord with the figures given in Chapter 3 for the first six months of the last embargo. The last embargo was much more costly overall because the price increases during the embargo persisted well after the embargo was over. A recurrence of this for future embargoes, however, is not anticipated for reasons given in Chapter 3.

The Cost of a Strategic Reserve

Storage costs were estimated using $1.23/bbl capital cost, maintenance cost of $0.0141/bbl, and a cost of purchasing oil which is either $7 a barrel or $11 a barrel, depending on the scenario. These were distributed over time as reflected in the Table A-1. The timing is based on some judgments about leaching water rates, deep-water port construction and the availability of tankers. These considerations suggest a fill rate of 183 million barrels a year or about 500,000 barrels a day starting in 1978. The drawdown rate during an embargo is assumed to use up 75 percent of the storage at the end of 182 days.

The Cost of Conservation

Both the Ford program and the bills considered in the House of Representatives have envisioned various conservation measures. These also have their concomitant costs. Because they cause no additional income transfers to the oil exporting nations, the cost to the nation as a whole consists solely of the deadweight loss to energy users resulting from foregone consumption.

The only complication in performing this calculation lies in insuring that the correct demand curve is used to measure the deadweight loss. Embargo costs are always computed using the short-run demand curve (i.e., a demand curve exhibiting a low price elasticity of demand) because embargoes are not continual and anticipated. Conservation measures, however, such as gasoline taxes or crude oil tariffs are both continual and anticipated. Therefore, they operate on the long-run demand curve. In practice these calculations are accomplished using the procedures specified above with two modifications: (1) the price elasticities used for each year increase over time so that they equal the long-run elasticity by 1985, and (2) equation (A.8) is modified so that only the deadweight loss is captured. This last point implies mathematically that the term $(S_{tj} + Q_{tj} - I_{tj})$ is

_____
[b]The calculation is as follows:

$$\frac{22.85}{0.89}(11.65^{0.89} - 4.0^{0.89}) - (11.65 - 4.0)(10.7) = 68.32$$

million dollars per day or $12.3 billion over a 180-day period.

Table A-1

Temporal Pattern of Costs and Oil Availability from a Billion Barrel Storage Program

(Billions of 1975 Dollars)

	Capital Cost	Maintenance Cost	Oil in Place Million of Barrels	$11 Crude		$7 Crude	
				Oil Purchase	Total Cost	Oil Purchase	Total Cost
1975	$0.1	$0.0	$ 0	$0.0	$0.1	$0.0	$0.1
1976	0.2	0.0	0	0.0	0.2	0.0	0.2
1977	0.3	0.0	0	0.0	0.3	0.0	0.3
1978	0.1	——	100	1.1	1.2	0.7	0.8
1979	0.1	——	283	2.0	2.1	1.3	1.4
1980	0.1	——	465	2.0	2.1	1.3	1.4
1981	0.1	0.01	648	2.0	2.1	1.3	1.4
1982	0.1	0.01	830	2.0	2.1	1.3	1.4
1983	0.1	0.01	1,000	1.9	2.0	1.2	1.3
1984	0.0	0.01	1,000	0.0	——	0.0	——
1985	0.0	0.01	1,000	0.0	——	0.0	——

Source: The capital cost estimates came from the Ad Hoc Committee on the Domestic and International Monetary Effect of Energy and Other Natural Resource Pricing of the House Committee on Banking and Currency, "Oil Imports and Energy Security: An Analysis of the Current Situation and Future Prospects," Committee Print, 93rd Cong., 2d sess., September 1974, p. 155. The other estimates were taken from J.P. Childress, "Construction, Logistics and Cost Considerations for a One Billion Barrel Crude Oil Stockpile," Federal Energy Administration Office of Quantitative Methods, discussion paper, no date.

replaced by the term Q_{tj}^c, where Q_{tj}^c is the level of consumption after the conservation action is taken.

The Present Value of Expected Cost

The final calculation takes these outcome costs for each policy, aggregates them, weights them by the likelihood they will be incurred and discounts them back to the present. Let C_{tj}^1 be the embargo costs associated with a given year and given OPEC pricing strategy and C_{tj}^2 and C_{tj}^3 be, respectively, the costs of enforced conservation and storage with $j = 1$ denoting the case in which the world crude oil price falls. Furthermore, let L_1 be the probability of OPEC causing a fall in the world price of crude in 1978 and L_2 be the probability of an embargo in any year. The expected present value of future costs (E) is then given as:

$$E = \sum_{t-1}^{11} \frac{[L_1(C_{t1}^2 + C_{t1}^3) + L_1 \cdot L_2 \cdot C_{t1}^1 + (1 - L_1)(C_{t2}^2 + C_{t2}^3) + (1 - L_1) \cdot L_2 \cdot C_{t2}^2]}{(1 + r)^{t-1}} \quad \text{(A.9)}$$

where r is the discount rate which is assumed to be 0.10. To provide a sensitivity analysis this calculation is performed for three values of L_1 (0.25, 0.50 and 0.75) and three values of L_2 (0.10, 0.20, and 0.40).

Notes

1. Kaj Areskoug, "U.S. Oil Import Quotas and National Income," *Southern Economic Journal* XXXVII (January 1971): 307-17, and J.C. Burrows and J.A. Domencich, *An Analysis of the United States Oil Import Quota* (Lexington, Massachusetts: Lexington Books, 1970).

2. Federal Energy Administration, *Monthly Energy Review*, March 1975, p. 57.

Appendix B:
Data

The key variables used in the analysis in Chapter 9 are presented in Table B-1.

Table B-1
Key Variables Used in the Calculation of Costs for Each Policy Package for Two OPEC Pricing Strategies, Selected Years

Variables and Policy	$7 Crude Price After 1978		$11 Crude Price Through 1985	
	1975	1985	1975	1985
U.S. consumption of products (Millions of barrels per day)				
Base case	17.0	23.3	17.0	19.0
HR6860	16.9	21.9	16.9	18.1
HR6860 with accelerated supply[a]	16.8	21.5	16.8	17.9
President's program without tariff	16.6	22.8	16.6	18.7
President's program with tariff	16.4	20.2	16.4	17.3
Imports of petroleum and petroleum products (Millions of barrels per day)				
Base case	6.3	11.9	6.3	3.4
HR6860	6.2	10.4	6.2	2.6
HR6860 with accelerated supply	6.1	5.9	6.1	2.3
President's program without tariff	5.8	7.3	5.8	1.4
President's program with tariff	5.6	4.7	5.6	0.0
Average petroleum product price ($1.975 per barrel)				
Base case[b]	8.44	7.00	8.44	11.00
HR6860	9.79	8.28	9.79	12.46
HR6860 with accelerated supply	9.79	8.28	9.79	12.46
President's program without tariff	11.00	7.00	11.00	11.00

Table B-1 (cont.)

Variables and Policy	$7 Crude Price After 1978		$11 Crude Price Through 1985	
	1975	1985	1975	1985
President's program with tariff	13.00	9.00	13.00	13.00
Price elasticity of demand in response to a 1975 price increase[c]	0.067	0.46	0.067	0.46

Source: The 1975 quantity numbers were generated by the FEA short-term forecasting model, which is documented in Federal Energy Administration, *National Petroleum Product Supply and Demand: October 1974 through November 8, 1974.* The 1985 estimates came from the FEA model. A description of this model can be found in Federal Energy Administration, *Project Independence Report* (November 1974), appendix pp. 195-282.

[a]The lower petroleum consumption levels for a program including an accelerated supply strategy results mainly from its component program to convert utilities from petroleum to coal insofar as possible.

[b]These prices assume that the amount of old oil declines linearly to zero in 1985.

[c]The relevant elasticities are assumed to be the same for all programs. The very low elasticity in the first year reflects the fact that the program would be in effect for only part of a year.

Index

Thomas H. Tietenberg, Assistant Professor of Economics at Williams College, received the Ph.d. in economics from the University of Wisconsin at Madison. During the year in which this book was written he was recipient of a Brookings Economic Policy Fellowship serving as Director of the Macroeconomic Impact Division in the Federal Energy Administration. Dr. Tietenberg has been a research associate in the John F. Kennedy School of Government at Harvard University and a member of a team of economists brought into the Defense Department to apply economic analysis to personnel planning. This research has led to one coauthored book, *The Automobile and the Regulation of Its Impact on the Environment*, published in 1975 and several articles in leading journals, such as the *Quarterly Journal of Economics, The American Economic Review* and *Public Policy*.